PHOTOCHEMISTRY
OF
MACROMOLECULES

PHOTOCHEMISTRY OF MACROMOLECULES

Proceedings of a Symposium held at the
Pacific Conference on Chemistry and Spectroscopy,
Anaheim, California, October 8-9, 1969

EDITED BY

RONALD F. REINISCH

NASA—Ames Research Center
Moffett Field, California

PLENUM PRESS • NEW YORK–LONDON • 1970

Library of Congress Catalog Card Number 70-127936
ISBN 978-1-4684-8037-5   ISBN 978-1-4684-8035-1 (eBook)
DOI 10.1007/978-1-4684-8035-1

© 1970 Plenum Press, New York
Softcover reprint of the hardcover 1st edition 1970
A Division of Plenum Publishing Corporation
227 West 17th Street, New York, N.Y. 10011

United Kingdom edition published by Plenum Press, London
A Division of Plenum Publishing Corporation, Ltd.
Donington House, 30 Norfolk Street, London W.C.2, England

PREFACE

Our knowledge of the photodegradation of polymers, chemical evolution, photosynthesis, visual perception and the biological effects of light depends heavily on our ability to elucidate the primary photochemical processes of macromolecules. This volume brings together for the first time from the fields of natural as well as synthetic polymers a group of reports dealing with macromolecular photochemistry. Since macromolecular photochemistry is an expanding new field that crosses the boundaries between classical disciplines, the reader will encounter the employment of diverse scientific approaches and unfamiliar terminology. However, it has become increasingly apparent that researchers in these fields have much to learn from each other. Although this book is not intended to give a detailed survey of the photochemistry of macromolecules, it does represent the editor's perspective on the relationship between theory, kinetic studies and the synthesis aspects of photochemistry.

The ideas expressed by the contributors offer a valuable composite of theoretical and experimental approaches for those who are concerned with problems which have photochemical relevance, and show that investigators from different fields share many concepts and perhaps some common problems. This novel array of present knowledge should provide a basis for organizing and understanding photochemical information from chemistry, physics, biology and medicine. While of particular value to the research worker, the book also should be of interest to the graduate student about to embark on a problem in macromolecular photochemistry.

It is a privilege to express my deepest appreciation to all those who made the symposium and this book possible. Many of the participants are acknowledged by their papers published in this volume but I am also indebted to Professor Mitchel Shen of the University of California, Berkeley and Professor John Heise of the Georgia Institute of Technology, who gave encouragement and support during the formative stages of this project.

Palo Alto, California   Ronald F. Reinisch
April 17, 1970

LIST OF CONTRIBUTORS

G. M. Androes, Ames Research Center, NASA, Moffett Field, California

A. Christopher, General Electric Research & Development Center, Schenectady, New York

W. B. Dandliker, Division of Biochemistry, Scripps Clinic & Research Foundation, La Jolla, California

E. M. Evleth, Division of Natural Sciences, University of California, Santa Cruz, California

A.K. Fritzsche, General Electric Research & Development Center, Schenectady, New York

H. R. Gloria, Ames Research Center, NASA, Moffett Field, California

J. E. Guillet, Department of Chemistry, University of Toronto, Toronto, Canada

A. V. Guzzo, Chemistry Department, University of Wyoming, Laramie Wyoming

M. Heskins, Department of Chemistry, University of Toronto, Toronto, Canada

H. H. G. Jellinek, Department of Chemistry, Clarkson College of Technology, Potsdam, New York

F. Kierszenbaum, Division of Biochemistry, Scripps Clinic & Research Foundation, La Jolla, California

J. F. Kryman, Eastman Kodak Company, Rochester, New York

C. B. Leman, Chemistry Department, University of Wyoming, Laramie, Wyoming

S. A. Levison, Division of Biochemistry, Scripps Clinic & Research Foundation, La Jolla, California

M. L. MacKnight, Department of Biology, University of Utah, Salt Lake City, Utah

A. D. McLaren, College of Agricultural Sciences, University of California, Berkeley, California

H. L. Needles, Department of Consumer Sciences, University of California, Davis, California

J. Pavlinec, Polymer Institute, Slovak Academy of Science, Bratislava, Czechoslovakia

R. Pecora, Department of Chemistry, Stanford University, Stanford, California

G. L. Pool, Chemistry Department, University of Wyoming, Laramie, Wyoming

R. O. Rahn, Biology Division, Oak Ridge National Laboratory, Oak Ridge, Tennessee

R. F. Reinisch, Ames Research Center, NASA, Moffett Field, California

K. C. Smith, Department of Radiology, Stanford University School of Medicine, Stanford, California

J. D. Spikes, Department of Biology, University of Utah, Salt Lake City, Utah

M. S. Toy, Douglas Advanced Research Laboratories, McDonnell Douglas Corporation, Huntington Beach, California

M. Weissbluth, Department of Applied Physics, Stanford University, Stanford, California

A. N. Wright, General Electric Research & Development Center, Schenectady, New York

CONTENTS

A BRIEF HISTORY OF THE PHOTOCHEMISTRY OF MACROMOLECULES

A. Douglas McLaren

College of Agricultural Sciences

University of California, Berkeley, 94720

INTRODUCTION

Ninety years ago Downes and Blunt (1) reported that the activity of invertase (zymase) can be destroyed by light and that irradiated solutions can lose residual activity on standing.* The significance of this loss of residual activity was not appreciated until much later (2) (3) (4) when it became clear that some enzyme molecules are unstabilized by ultraviolet radiation, UV; further, some in a population of radiation modified molecules are altered so as to have lost more activity toward one type of substrate than another (5) (6). A differential loss of activity doubtless parallels selective photochemical events in the molecule (7), for example, when one or more tryptophan residues are modified the proteolytic activity of trypsin is annililated, but the esterase activity is almost undiminished (8). To-date it is not clear whether some molecules have lost all activity toward one type or the other or whether molecules can have altered activity toward one type and unaltered activity toward the other. In any case, some molecules are so altered by irradiation as to lose activity almost at once and others to lose activity on standing, or warming, perhaps by

---

* Although a number of biopolymers, including cellulose, wool and rubber, have been irradiated with ultraviolet light, this outline will be confined to molecularly dispersed substances the irradiation of which can have marked biological consequences, either if irradiated pure in solution, if added, after irradiation, to living cells, or if irradiated within viable cells. These include enzymes, nucleic acids and viruses. Artificial polymers will be discussed by others.

rupture of hydrogen bonds or by oxidation and other changes leading simultaneously to absorbance changes (2).

With malt diatase it was found that the reciprocity law holds (9) and with pancreatin (10) and pepsin (11) first order inactivation was observed during irradiation. The reason for these kinetics has since been elucidated in phenomenological terms (12) but a mechanistic explanation is still incomplete. This latter point remains one of the important challenges. The question becomes, what are the photochemical and subsequent chemical changes which lead to enzyme inactivation? From an organic chemical point of view, almost all the expected, possible changes taking place were already observed by 1949, but which changes can be correlated directly with enzyme, E, radiation-inactivation still needs to be worked out.

PHOTOCHEMISTRY OF ENZYMES

Early literature has been summarized (12) and in the past two decades progress has been reviewed (13) (14) (7) (15). At one time it was suggested that inactivation studies as a function of wavelength might reveal curves similar in shape to those for model substances and thereby give some clue as to the chemical groupings about the active site (16). Eventually this was done with dry trypsin by Setlow and Doyle (17). They resolved their action spectrum into two components, one representing photons absorbed by cystine residues, the other representing photons absorbed by the remainder of the molecule. At 2537Å cystine residues are of paramount importance, but aromatic residues are more important at ca. 2850Å and even peptide bonds become important at short wavelengths.

If irradiation can lead to inactivation by more than one chemical mechanism, then the total rate of inactivation will be given by (provided there is no energy transfer among residues):

$$-d(E)/dt = \frac{I_{abs}}{\epsilon_E[E_o]} \sum^{ij} n_{ij} \epsilon_{ij} [E] \phi_{ij} \qquad 1)$$

where [E] is the concentration of enzyme, with an initial concentration $[E_o]$ and an extinction coefficient ϵ_E, n_i is the number of chromophores of kind i in the molecule, having residue extinction coefficient ϵ_i and undergoing photochemical reaction for modification with a quantum yield ϕ_i. The kinds of amino acid residues of photochemical significance in the enzyme include aromatic groups, and cystine if present. If n_i is two or more and a given kind of residue, for example, cystine, has a different ϕ_i and ϵ_i for each position in a molecule, then n_{ij} residues with specific ϵ_{ij} and ϕ_{ij}

must be used (18) (19). The residues n_{ij} may be less than the total for each kind in a molecule if the photochemical alteration of one or more of them is of no consequence to enzyme activity (7). If hydrogen bonding can be altered by radiation, leading to denaturation-inactivation of the enzyme (3) (20), either during irradiation or on standing after irradiation, an additional term must be added to equation 1. This is merely a summation equation: it should account for the overall rate and quantum yield,

$$\Phi = \frac{\sum_{}^{ij} n_{ij} \epsilon_{ij} \phi_{ij}}{\epsilon_E} \qquad\qquad 2)$$

provided one has the correct empirical values for n, ϵ, and ϕ for each of the vital residues. Equation 2 can have predictive value if and when all values of ϵ_{ij} are nearly the same and equal to ϵ_i for a particular kind of residue and for ϕ_{ij} only one value, ϕ_i is observed for this kind of residue. It should be noted that ϕ_i or ϵ_i in equation 1 may or may not be close to the value of free amino acid alteration or absorbance respectively. Before the availability of empirical values for each residue in each enzyme, values for free amino acids were simply used to test equation 1, Table 1, (21) (19) (22).

Table 1

Some calculated and known quantum yields for enzyme inactivation by ultraviolet radiation (2537Å)

Enzyme	Chromophores	Quantum yield		Found
		Calculated		
		Eq. 1	Eq. 2	
Carboxypeptidase	$Cys_0 \cdot His_8 \cdot Phe_{15} \cdot Try_8 \cdot Tyr_{19}$	0.01	-	0.005
Subtilisin A	$Cys_0 \cdot His_5 \cdot Phe_4 \cdot Try_1 \cdot Tyr_{13}$	0.005	-	0.007
Trypsin	$Cys_6 \cdot His_1 \cdot Phe_3 \cdot Try_4 \cdot Tyr_4$	0.02	-	0.015
	(Cystine only)	0.014	0.014	
Ribonuclease	$Cys_4 \cdot His_4 \cdot Phe_3 \cdot Try_0 \cdot Tyr_6$	0.03	-	0.027
	(Cystine only)	0.026	0.035	
Lysozyme	$Cys_5 \cdot His_1 \cdot Phe_3 \cdot Try_8 \cdot Tyr_2$	0.014	-	0.024
	(Cystine only)	0.010	0.020	

As may be seen from Table 1, calculated values are only roughly equal to found values and the possible reasons for lack of good agreement are legion. Φ are usually dependent on pH [perhaps more so than ϕ_i for a free amino acid (23)] for unknown reasons; different values for a given residue are known [for example, cystine residues have different ϵ_{ij} and ϕ_{ij} in trypsin and ribonuclease (24) (25)]; the integrity of each of the cystine residues are not of equal importance; energy transfer among the residues, leading to photochemical action beyond the site of absorption, is possible; etc. Cystine residues, if present, may be the most sensitive or "weak-link". This has lead to the notion that there are two opposing points of view of enzyme inactivation (26), namely the "weak-link" hypothesis (27) and that embodied in equation 2. This is misleading since the notion partly depends on whether or not cystine disruption (if cystine residues are present) is or is not of paramount importance in inactivation and partly on whether any energy transfer of photochemical interest is possible. The equation

$$\Phi = \sum^{i} n_i \epsilon_i \phi_i / \epsilon_E \qquad \qquad 2a)$$

has only the following experimental photochemical quantities: Φ, ϵ_E, ϵ_i, and ϕ_i. That is, n_i, the number of residues of a given type that are essential for activity cannot be directly determinable, even if all residues of a given amino acid type have equal sensitivities ϕ_i and absorbancies ϵ_i. If e.g., two types of residues are important

$$\Phi = n_{ss} \epsilon_{ss} \phi_{ss} / \epsilon_E + n_{Try} \epsilon_{Try} \phi_{Try} / \epsilon_E \qquad 2b)$$

we have one equation and two unknowns. One extreme but tractable approach is to assume either that all n_i groups are necessary, or to specify n_i from independent studies of the essentiality of a fraction α. This was done by McLaren and Shugar. Another extreme approach is to ignore all residues but cystine, if present.

Both the latter, adopted by Augenstein, and the former require a term for "denaturation" on hydrogen bond breakage etc. as necessary to obtain a satisfying result. Perrase et al. have adopted a method for finding n_i; it involves an evaluation of equation 2b at two wavelengths, with corresponding values of ϵ_i, ϕ_i and Φ and solving the equations simultaneously for n_{ss} and n_{Try}. [Rathinasamy and Augenstein (28) have also obtained data at more than one wavelength, which can be interpreted to mean that at some wavelengths energy transfer from a UV absorbing chromophore, followed by -ss- photolysis is highly probable.]

Dose (19) has taken into account energy transfer among enzyme residues by writing

$$\phi = \alpha \ \phi^o_{s-s} \qquad\qquad\qquad 3)$$

where α is the fraction of total cystine residues essential for enzyme activity and ϕ^o_{s-s} is an averaged cystine photochemical sensitivity.

$$\phi^o_{s-s} = f_1 \phi_{s-s} + f_2 \phi'_{s-s} \qquad\qquad\qquad 4)$$

when f_1 is the fraction of light absorbed by cystine residues, f_2 is the fraction of light absorbed by phenylalanine and tyrosine residues, ϕ_{s-s} is the quantum yield for destruction of a cystine residue by direct absorption and ϕ'_{s-s} is a quantum yield which accounts for an increase in quantum yield for cystine destruction by light absorbed by tyrosine and phenylalanine residues. The calculated value for ribonuclease (which lacks tryptophan) becomes 0.035 by equation 4, 0.014 for trypsin and 0.020 for lysozyme. These are in somewhat better agreement with found values for the enzymes containing cystine.

With these equations as guidelines we turn to specific cases. Carboxypeptidase has received considerable attention. It lacks cystine and therefore one can look directly at the involvement of aromatic residues in photoinactivation. A selective destruction of tryptophanyl and tyrosyl residues has been observed (29), along with a slight decrease in the helical content; the latter indicating denaturation of the enzyme molecules. Irradiation liberates zinc and decreases both esterase and peptidase activities. As tyrosyl residues are destroyed, esterase activity increases, however. With increasing doses tryptophan residues are also destroyed (30). It was concluded that radiation exerts its effects on the function of carboxypeptidase both by destruction of specific amino acid residues and through profound alteration of the structure as evident from the physiochemical characteristics of the irradiated enzyme (31). The latter contained three species, two of which were denatured. The non-denatured product had altered esterase activity and one destroyed tyrosyl residue; tryptophan and histidine were unaltered. The enzymatic changes in the denatured fractions were nonspecific since both esterase and peptidase activities decreased in parallel fashion. In summary, at least three distinct processes are involved in the overall photoinactivation of carboxypeptidase, namely loss of zinc ($\phi = 1.6 \times 10^{-3}$), alteration of tyrosine at the active center ($\phi = 5 \times 10^{-4}$) and denaturation ($\phi = 2.8 \times 10^{-3}$). Piras and Vallee (22) concluded that "the quantum yield of inactivation of an enzyme can only be predicted exactly when the nature and number of amino acid residues directly or indirectly involved in enzymatic activity can be established". The sum of the quantum yields observed for the three processes is smaller than the predicted value

(Table 1), confirming that only some of the absorbing amino acids
are essential for activity. Finally, it was observed that a modification of two or three tryptophanyl residues could account for
the quantum yield for denaturation, ϕ_{den}.

Originally equation 1 lacked a term for denaturation or hydrogen bond breakage, although the need was obvious and was based on
earlier observations involving the denaturation of chymotrypsin (3)
(20) and trypsin (32).

Most of the other enzymes studied in detail contain one or
more cystine residues and the photolysis of these residues is very
important at 2537Å (12) (13) (17) (27). Rathinasamy and Augenstein
do not accept equation 1 in its original form (21) since in this
form it embodied the assumptions that all cystine residues were of
equal importance and/or had equal quantum yields for photolysis.
This now seems true for insulin, but is not general. However, many
publications have shown that cystine disruption is correlated with
the UV inactivation of enzymes (13) (25) and that the value of both
ϵ_i and ϕ_i for amino acid residues depend on the type of neighboring
residues as well as on the conformation of a protein molecule (21)
(29) (24). It has been concluded that there is an appreciable difference in the interactions between constituent aromatics and cystines in different proteins (28). In other words, equation 1 is
more of a summation equation for results and less a precise predictive equation than is desirable for any given enzyme. Equation 3
may be only appropriate for a given wavelength and set of experimental conditions: a severe test comes with a variation in wavelength of active UV. Cystine quantum yields in RNAase are similar
at 2650 and 2540Å, whereat the fraction of the light absorbed by
cystine residues is relatively large. At 2290Å the yield for cystine photolysis is greater than expected from quanta absorbed by
cystine, i.e., approximately 85% of the yield observed must reflect
the effects of photons absorbed elsewhere in the molecule. At least
one other inactivation step is involved with RNAase: no tyrosine
is lost at 2537Å and 2650Å but at 2290Å the number of tyrosines lost
per molecule of RNAase inactivated is close to 1.0. In conclusion,
at a given wavelength, various cystines can be disrupted at different rates; the most photo-sensitive cystine residues are not necessarily those critical for activity and at some other wavelength the
relative importance of cystine and tyrosine destruction, leading to
enzyme inactivation, can be different. Further, during inactivation
of an enzyme, amino acid residues are not destroyed by a process as
random as that suggested by equation 1 in its original form (33).

Excited state studies point to an interaction between excited
tyrosyl residues and vicinal disulfide linkages in ribonuclease (34).
When free cystine is irradiated in the presence of free tyrosine,
the quantum yield for cystine destruction is increased by 15-20%
(6). Perrase et al., find that the destruction of cystine in

trypsin also occurs through energy absorbed by aromatic residues. Thus the notion of photochemical significant energy transfers in photoinactivation of enzymes seems well established and represents one of the most significant advances in this field.

At the active site of lysozyme, cystine residue no. 64 is linked to tryptophan residue nos. 62 and 63; these are both involved in substrate binding (35). Although Ferrini and Zito could not find altered tryptophan in irradiated lysozyme, they reported an altered histidine residue (36). These results have been totally refuted experimentally by Churchich (37) and theoretically (38) (28). Churchich found a destruction of one tryptophan per enzyme molecule inactivated, formation of about one SH group, and a decrease in the fluorescence quantum yield for the enzyme, in general agreement with the work of Luse and McLaren and of Dose (39). This is not a proof that some molecules are inactivated by reduction of -SS- bonds and some by tryptophan destruction as -SS- may be reduced in a molecule simultaneously with oxidation of tryptophan (40). An effort should be made to fractionate irradiated lysozyme to see if two or more inactive kinds of molecules are present in the product. Incidentally, other products besides -SH are found after UV irradiation of -SS- compound. So one may not expect a one-to-one correlation of photolysis of RSSR to RSH and enzyme inactivation in the general case (Forbes and Savige, (41) and subsequent papers).

In ribonuclease, trypsin, and lysozyme the most photosensitive cystine residues are not the ones most vital to enzyme activity (24). Loss of ribonuclease activity requires that at least half of the constituent cystine residues be disrupted, Table 2. With trypsin and lysozyme the rate of cystine destruction is also greater than the initial rate at which activity is lost and there is considerable tryptophan destruction in both molecules.

Table 2

The relation between loss in enzyme activity and cystine disruption or tryptophan destruction for quanta absorbed anywhere in the protein (Risi et al., 1967).

	n_{s-s}	Φ_{enzyme}	ϕ_{s-s}/Φ	ϕ_{Try}/Φ
Ribonuclease	4	0.031	2.4	-
Lysozyme	4	0.014	2.5	2.5
Trypsin	6	0.020	2.5	1.8

PHOTOCHEMISTRY OF RIBOSENUCLEIC ACIDS

   The most extensively studied RNA is that from tobacco mosaic
virus and it will be discussed within a framework of the photochemistry of this oldest known virus. Tobacco mosaic virus (TMV) is
historically unique. According to Mulvania (42), it is the first
of the so-called filterable viruses. Mulvania found that the sensitivity of TMV toward ultraviolet radiation (u.v.) was relatively
low, more like that of an enzyme, than of a bacteria. In fact,
Smith (43) found that Bacillus prodigiosus was sixty times more
sensitive than TMV and concluded, therefore, that the casual agent
(TMV) is not an organism.

   "Killing" of crystalline TMV was found to be an exponential
function of UV dose and to be faster when pure than in plant juice.
The situation, it was concluded, obtained because of an inner filter
effect by absorbing constituents in juice (44). That virus inactivation by u.v. might have much in common with gene mutation, as in
Drosophila, was also suggested by Price and Gowan; efforts to induce
mutation in TMV with irradiation have still not born fruit (45).
Preliminary action spectra for killing bacteria, coagulation of
protein and inactivation of TMV were reported by Duggar and Hollaender in 1934 (46), and it was stated that they "are of the same
general type". Further, resistance of TMV to u.v. is greater than
that of bacterial spores of Bacillus subtilis and B. megatherium
(47).

   Rough calculations by Uber (48) indicated that the quantum
yield for inactivation (Φ) for this, by now a well-established "one-
hit" process, was lower than that for enzyme inactivation, and he
suggested that the higher the molecular weight (M) of the bio-
particle the lower the quantum yield. Sometime later it became
clear that in order to truly compare sensitivities of bio-particles,
the absorption characteristics must also be taken into account and
that the quantum yields for enzymes (several), viruses (two) and a
bacterium were roughly inversely proportional to particle weights
to the two-thirds power (49).

   Hollaender and Duggar (50) used the modern photobiological
methods of Warburg in their classic paper on an action spectrum for
inactivation of purified virus, just after pure virus was partially
characterized by Stanley (51). In keeping with beliefs of the time
they stated that the relative absorption curve should have the
shape of the reciprocal of the inactivation curve if all the absorbed energy were used in the inactivation of the virus. In modern terms it is only essential that Φ be independent of wavelength.
Light scatter is also known to be great with TMV, but at that time
the scatter had not been measured. The features of their action
spectrum are still a challenge (52). A shoulder in the spectrum

at 2600Å might have suggested nucleic acid as an absorber, but the sudden rise below 2480Å was mysterious.

After a molecular weight had been established for TMV it was possible to obtain a quantum yield, namely 4.3×10^{-5} (53), and still later it was found that the value increases by a factor of less than three from 2804 to 2483Å, with about a two-fold jump at 2300Å. Thus, unlike the infectious free RNA from TMV, with a nearly constant quantum yield, the absorption and inactivation (action) spectra cannot coincide (7). Any changes that may occur in virus protein during irradiation do not seem to contribute to loss of infectivity of TMV, although the protein coat is not without influence; while inside the virus, RNA is about 24 times more sensitive to inactivation at 2300 than at 2800Å. Inside, the virus RNA seems to be protected somewhat by the protein from inactivation damage at 2800 whereas at 2300Å the RNA is about as sensitive as when free (52).

UV inactivated TMV is not photoreversed by visible light on plant leaves, whereas inactivated RNA is (54). Tobacco gives somewhat more photorecovery than Chenopodium or Pinto bean under white light (55). With the bean, the greatest recovery (about 40 percent) has been found with "black light" at about 3650Å (56). Thus, u.v. at any wavelength causes at least two kinds of damage to free RNA, one of which is photorecoverable, but the photoreversible damage does not occur in the RNA irradiated inside the virus, even if it is subsequently isolated and assayed free on leaves (54). Percent photorecovery of inactivated RNA depends on the wavelength, and is greater following inactivation at 3020Å than at 2537, which also suggests two possible kinds of damage (57).

Stanley (51) reported that u.v. treatment of TMV produced an inactive product which retained its solubility and serological properties. The product, however, is more easily denatured by heat (53, 58). Prolonged irradiation leads to gaps in the walls of the virus particles, extending from the edge to the center of the particles (59) and eventually to loss of antigenicity (60, 61) and structure (62).

These degradative reactions take place with quantum yields considerably less than that for inactivation, however. Similarly, free RNA is inactivated with doses much less than required for depolymerization (63) and reconstitution of inactivated RNA with viral protein to give virus-like rods is possible (64). Further, the intrinsic viscosity of RNA extracted from TMV after inactivation by u.v. does not differ from RNA extracted from non-irradiated TMV (65).

Irradiation of TMV in D_2O or in H_2O gives the same quantum

yield (66). The supposition that water is not involved in TMV inactivation is not tenable, however, since uridine hydrate has been
found in RNA extracted from irradiated TMV (67). Since TMV A-protein
loses water on polymerizing to rods (68), it seems likely that the
TMV is less hydrated, i.e., is surrounded by less "free-water" than
when free in solution. Further,the configuration of RNA within the
virus (69) is quite different from that of free RNA in solution (70),
wherein, secondary and tertiary structure can depend on ionic
strength (71). These and other ideas need to be explored.

   Attempts to identify the nature of the primary photochemical
reactions involved in inactivation of TMV or the RNA have been sin-
gularly without success. Although some peptide remains attached to
RNA following extraction from irradiated TMV under certain conditions
(72), this association reaction seems not directly involved (52).
Pyrimidine dimer formation occurs during irradiation of RNA, but no
simple correlation was shown with the biologically defined lesion
(73). Pyrimidine hydrate formation also takes place during irradiation of RNA (74) which is not surprising since inactivation is faster
in H_2O than in D_2O at low ionic strength (66), an obvious isotope
effect. Calculations based on optical changes during irradiation
suggest that only a few bases are altered during inactivation (63),
but the causal relationship is still obscure. The relationship may
be indirect and analogous to that found for transfer RNA; with tRNA
anticodon sites do not seem to be involved, rather u.v. damage produces a change in the conformation which in turn inactivates the
tRNA (75).

   Although we know that free RNA undergoes at least two kinds of
photoinactivation, the role of coat protein in preventing one of
these is also unknown: RNA extracted from inactive TMV cannot be
photoreactivated and RNA inactivated in the free state and then re-
polymerized with protein cannot be photoreactivated. Upon re-
extraction from reconstituted, inactive TMV, the RNA can, however,
be photoreactivated to the almost usual degree (76).

   Incidentally, quantum yield calculated for TMV-protein by means
of equation (2) above (38), namely 0.0047, may be compared with
0.005 based on loss of antigenicity as measured by Kleczkowski (60).

                PHOTOCHEMISTRY OF DESOXYRIBOSE NUCLEIC ACIDS

   Serious work on the photochemistry of sodium thymonucleate was
first reported by Hollaender et al. (77) who observed changes in
gross properties such as a decrease in solution viscosity. That
loss of biological activity of transforming-DNA involved more subtle
changes, without reduction in viscosity, was described by Zamenhof
et al.(78). A large advance came with the work of Rupert (79) who

showed that the damage could, in part, be reversed both in vivo and in vitro under suitable conditions and with the aid of photo-recovery enzyme systems. The formation of pyrimidine hydrates and dimers, as described above, is directly involved in inactivation of these macro-molecules and further discussion will be left to Dr. R. O. Rahn later in this symposium.

Acknowledgements - This work was supported in part by the U. S. Atomic Energy Commission Contract AT(11-1)-34, Project 116.

SUMMARY

Important observations leading to understanding of the photo-chemistry of enzymes and nucleic acids are discussed. A pseudo-first order equation describes rates of inactivation fairly well but any general equation will be an over simplification. Enzyme inactiva-tion involves reaction of many kinds of bonds, some of which may be broken in the primary process of photon absorption, some following energy transfer, and some thermally, perhaps hydrogen bonds. A one-to-one correlation of bonds broken and protein-inactivation may hold for some.

A history of the photochemistry of the first known virus (tobac-co mosaic) is presented. Hollaender and Duggar recognized that the virus contains nucleic acid because of the shape of their action spectrum for photoinactivation. An analysis of this spectrum in detail presents a challenge to photochemists because of the combined role of protein and nucleic acid in determining rates of inactiva-tion. At present these rates can not be explained simply on the basis of the photochemistry of each of the constituents; that is, the photochemistry of free infectious nucleic acid differs in sev-eral respects from that exhibited by nucleic acid irradiated in intact virus. At present it is believed that pyrimidine hydrate and dimer formations are important during photoinactivation.

REFERENCES

1. A. Downes and T. P. Blunt, Proc. Roy. Soc. London, 26, 199 (1879).
2. E. B. Sanigar, L. E. Krejci, and E. O. Kraemer, Biochem. J., 33, 1 (1939).
3. A. D. McLaren and P. Finkelstein, J. Am. Chem. Soc. 72, 5423 (1950).
4. L. G. Augenstein, C. A. Ghiron, K. L. Christ, and R. Mason, Proc. Natl. Acad. Sci. 47, 1733 (1961).
5. L. Augenstein, Science 129, 718 (1959).
6. K. Dose and G. Krause, Photochem. Photobiol. 7, 503 (1968).

7.   A. D. McLaren and D. Shugar, Photochemistry of Proteins and
     Nucleic Acids, Pergamon Press, Oxford, 1964.
8.   M. A. Coletti-Previero, A.Previero and E. Zuckerkandl, J. Md.
     Biol. 39, 493 (1969).
9.   A. H. Hutchinson and M. R. Ashton, Can. J. Res. 9, 49 (1933).
10. W. R. Thompson and R. G. Hussey, J. Gen. Physiol. 15, 9 (1931).
11. R. G. Hussey and W. R. Thompson, J. Gen. Physiol. 9, 217 (1925-
     1926).
12. A. D. McLaren, Adv. Enzymology 9, 75 (1949).
13. A. D. McLaren, Enzymologia, 18, 81 (1957).
14. L. Augenstein, Adv. Enzymology 24, 359 (1962).
15. J. K. Setlow, in Comprehensive Biochemistry, Vol. 27, M.
     Florkin and E. H. Stotz, Eds., Elsevier, New York, 1967.
16. A. D. McLaren, J. Polymer Sci., 1, 107 (1947).
17. R. Setlow and B. Doyle, Biochim. Biophys. Acta 24, 27 (1957).
18. A. D. McLaren, Abstracts 2nd International Biophysical Congress,
     Vienna, September, 1966, No. 573.
19. K. Dose, Photochem. Photobiol. 6, 437 (1967).
20. A. D. McLaren and R. A. Luse, Science, 134, 836, 1410 (1961).
21. R. A. Luse and A. D. McLaren, Photochem. Photobiol. 2, 343
     (1963).
22. R. Piras and B. L. Vallee, Biochem. 6, 2269 (1967).
23. I. H. Leaver and F. G. Lennox, Photochem. Photobiol. 4, 491
     (1965).
24. S. Risi, K. Dose, T. K. Rathinasamy, and L. Augenstein, Photo-
     chem. Photobiol. 6, 423 (1967).
25. L. Augenstein and P. Riley, Photochem. Photobiol. 3, 353 (1964).
26. N. I. Perrase, N. V. Kondakova, T. N. Kalabukhova, Y. A.
     Vladimirov, and L. K. Eidus, Biofizika 13, 24 (1968).
27. L. Augenstein and B. R. Ray, J. Phys. Chem. 61, 1385 (1957).
28. T. K. Rathinasamy and L. G. Augenstein, Biophys. J. 8, 1275
     (1968).
29. H. Fujioka and K. Imahori, J. Biochem. 53, 341 (1963).
30. R. Piras and B. L. Vallee, Biochem. 5, 849 (1966a).
31. R. Piras and B. L. Vallee, Biochem. 5, 855 (1966b).
32. L. Augenstein and C. A. Ghiron, Proc. Natl. Acad. Sci., U. S.
     47, 1530 (1961).
33. K. L. Grist, T. Taylor, and L. Augenstein, Rad. Res. 26, 198
     (1965).
34. J. W. Longworth, Photochem. Photobiol. 7, 587 (1968).
35. D. G. Smyth, W. H. Stein, and S. Moore. J. Biol. Chem., 238,
     227 (1963).
36. V. Ferrini and R. Zito, J. Biol. Chem. 238, PC 3824 (1963).
37. J. E. Churchich, Biochim. Biophys. Acta, 126, 606 (1966).
38. A. D. McLaren and O. Hidalgo-Salvatierra, Photochem. Photobiol.
     3, 349 (1964).
39. K. Dose, Biophysik 1, 316 (1964).
40. K. Dose, Photochem. Photobiol. 7, 671 (1968).
41. W. F. Forbes and W. E. Savige, Photochem. Photobiol. 1, 1
     (1962).

42. M. Mulvania, Phytopath. 16, 853 (1926).
43. F. F. Smith, Ann. Missouri Bot. Garden 13, 425 (1926).
44. W. C. Price and J. W. Gowen, Phytopath. 27, 267 (1927).
45. A. Siegel, Adv. Virus Res. 11, 25 (1965).
46. B. M. Duggar and A. Hollaender, J. Bact. 27, 219 (1934).
47. B. M. Duggar and A. Hollaender, J. Bact. 27, 241 (1934).
48. F. M. Uber, Nature 147, 148 (1941).
49. A. D. McLaren, Acta. Chem. Scand. 4, 386 (1950).
50. A. Hollaender and B. M. Duggar, Proc. Natl. Acad. Sci. U. S.
     22, 19 (1936).
51. W. M. Stanley, Science 81, 644 (1935); Science 83, 626 (1936).
52. A. Kleczkowski and A. D. McLaren, J. Gen. Virol. 1, 441 (1967).
53. G. Oster and A. D. McLaren, J. Gen. Physiol. 33, 215 (1950).
54. F. C. Bawden and A. Kleczkowski, Nature 183, 503 (1959).
55. H. Werbin, O. Hidalgo-Salvatierra, J. Seear and A. D. McLaren,
     Virology 28, 202 (1966).
56. O. Hidalgo-Salvatierra and A. D. McLaren, Photochem. Photobiol.
     9, 417 (1969).
57. V. Merriam and M. P. Gordon, Proc. Natl. Acad. Sci. U. S. 54,
     1261 (1965).
58. A. Kleczkowski, Biochem. J. 56, 345 (1954).
59. A. D. McLaren and A. Kleczkowski, J. Gen. Virol. 1, 391 (1967).
60. A. Kleczkowski, Photochem. Photobiol. 1, 291 (1962).
61. Y. Miyamoto, J. Virol. 4, 256 (1954).
62. A. H. Zech, Z. Naturf. 16b, 520 (1961).
63. A. D. McLaren and W. N. Takahashi, Rad. Res. 6, 532 (1957).
64. H. Fraenkel-Conrat, M. Staehelin and L. V. Crawford, Proc. Soc.
     Exper. Biol. Med. 102, 118 (1959).
65. A. Buzzell, D. T. Trkula and M. A. Lauffer, Proc. First Nat.
     Biophysics Conference, Columbus, 1957, pp. 254-255. Yale
     University Press, New Haven (1959).
66. M. Tao, M. P. Gordon, and E. W. Nester, Biochem. 5, 4146 (1966).
67. M. Tao, G. D. Small and M. P. Gordon, J. Mol. Biol. 38, 75
     (1968).
68. M. A. Lauffer, Biochem. 3, 731 (1964).
69. E. M. Schachter, I. J. Bendet and M. A. Lauffer, J. Mol. Biol.
     22, 165 (1966).
70. C. A. Knight, In Plant Virology, p. 292. University of Florida
     Press (1964).
71. D. W. McMullen, S. R. Jaskunas and I. Tinoco, Biopolymers 5,
     589 (1967).
72. J. Goddard, D. Streeter, C. Weber and M. P. Gordon, Photochem.
     Photobiol. 5, 213 (1966).
73. V. Merriam and M. P. Gordon, Photochem. Photobiol. 6, 309
     (1967).
74. G. D. Small, M. Tao and M. P. Gordon, J. Mol. Biol. In press.
75. P. S. Sarin and H. E. Johns, Photochem. Photobiol. 7, 203
     (1968).
76. G. D. Small and M. P. Gordon, Photochem. Photobiol. 6, 303
     (1967).

77. A. Hollaender, J. P. Greenstein and W. V. Jenrette, J. Nat.
        Canc. Inst. $\underline{2}$, 23 (1941).
78. S. Zamenhof, G. Leidy, E. Hahn, and H. E. Alexander, J.
        Bacteriol. $\underline{72}$, 1 (1956).
79. C. S. Rupert, J. Gen. Physiol. $\underline{45}$, 703 (1962).

PHYSICAL AND ENVIRONMENTAL FACTORS INFLUENCING THE PHOTOCHEMISTRY OF DNA*

R. O. Rahn

Biology Division, Oak Ridge National Laboratory,

Oak Ridge, Tennessee

INTRODUCTION

Deoxyribonucleic acid (DNA) contains four bases — adenine, guanine, thymine, and cytosine — each of which absorbs strongly in the ultraviolet (UV), with a maximum at ~260 nm. Irradiation of DNA in solution with UV light leads to the formation of cyclobutane dimers of the type

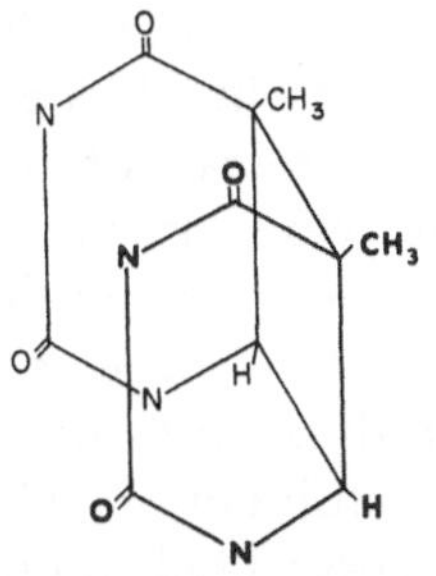

between adjacent thymines. Cytosine–cytosine dimers and thymine–cytosine dimers are also produced by UV irradiation, though to a lesser extent. The purines, adenine and guanine, are considerably less reactive to UV than the pyrimidines. Thymine dimers were first obtained by Beukers and Berends from irradiated frozen solutions of thymine (1) and from irradiated DNA (2), and later by Wacker et al. (3) from irradiated vegetative bacterial cells. Subsequently Setlow et al. showed that such dimers inhibit the synthesis of DNA in growing bacteria (4). Several review articles (5, 6, 7) are available, discussing both the photochemistry of DNA and the nature and biological significance of pyrimidine dimers.

*Research sponsored by the U.S. Atomic Energy Commission under contract with the Union Carbide Corporation

Irradiation of bacterial spores results in the formation of a thymine-
derived photoproduct (spore photoproduct) not normally observed in either
irradiated vegetative cells or in isolated DNA (8).   Since little or no thymine
dimerization occurs in spores, it appears that the physical state of DNA is
considerably different in spores than it is in vegetative cells or in solution.
The structure of the spore photoproduct is not known but it is believed to involve two thymine residues which may be connected via an azetane ring as
follows (9)

Of interest are the structures of DNA that favor this photoproduct and the
environmental factors that influence its formation.

In this paper our attention will be directed towards formation of the thymine dimer and the spore photoproduct in DNA and poly dT.   We shall examine
in particular the influence of the physical state or the environment of the DNA
upon the facility with which these two products are formed.   Such studies help
in understanding the influence structure has on the mechanism of photochemical
reactions that occur in macromolecules.   The techniques described here for
controlling the photochemical yields by varying experimental conditions such
as temperature, relative humidity, solvent, or pH seem applicable to most
polymer systems and may in some cases lead to unusual photoproducts.

THYMINE DIMERIZATION

Thymine dimers are formed in polynucleotides that contain adjacent thymine residues.   This situation exists in the dinucleotide TpT and the polynucleotides poly dT, poly (dA·dT), and DNA, but does not exist in poly d(A-T),
where the thymine and adenine residues alternate on a given strand.   Removal
of the adenine residues of poly d(A-T) by heating in acid (depurination) gives
a polymer in which dimerization occurs between the undisturbed thymine residues such that at high doses more than 50% of the thymine is dimerized (10).
Similarly, in depurinated DNA the yield of dimers is twice that obtained normally for native DNA (10, 11).   Thymines, normally separated by either
adenine or guanine, presumably become adjacent following depurination and
these then are free to dimerize.

Absorption of a photon by the cyclobutane ring leads to a splitting of the thymine dimer. Since the dimer absorbs to shorter wavelengths than the monomer, the final steady-state concentration of thymine dimers in a polynucleotide is a function of the wavelength of irradiation. That is, high yields of dimer are obtainable only at long wavelengths of irradiation, where the dimer absorption is not appreciable. For example, in Escherichia coli DNA, 25% of the bases are thymine. Hence, if we assume a random distribution of bases, there is a 50% probability that a thymine will have a thymine neighbor on either side available for dimerization. It is difficult, however, to achieve more than 20% conversion of thymine to dimer with irradiation wavelengths of ~280 nm. Presumably, the dimer still absorbs enough at this wavelength to affect appreciably the photoinduced steady-state equilibrium. Much higher yields of dimer (37%) were achieved by Lamola (12), who employed acetophenone as a triplet sensitizer and irradiated at wavelengths greater than 300 nm.

STACKING

A major factor that influences the dimerization of two adjacent thymines in a polynucleotide is the strength of the mutual interaction for stacking energy between the two bases (13). The importance of this stacking energy in dimerization was first demonstrated experimentally by Wacker and Lodeman (14); they measured the dimerization of TpT in a series of solvents which varied in their ability to unstack the dinucleotide. As shown in Table 1, there is a very good correlation between the yield of thymine dimer and the free energy difference $\Delta F°$ between the unstacked and stacked forms of TpT in these solvents. The yield of dimers is lowest in those solvents which do not favor stacking.

TEMPERATURE

T > 25°C

Since increasing the temperature of a polynucleotide will ultimately overcome the forces maintaining a stacked structure, we should expect to see a decrease in dimerization as the temperature is raised. Such an effect in TpT was observed by Zavil'gel'skii et al. (15), who showed that at 76°C very little dimerization occurred at all. At this temperature the bases in TpT are randomly arranged, as determined by CD measurements (15). Similar results were obtained by Tramer et al. (16) with poly rT. They found that the rate of dimerization decreased going from the ordered polymer (T < 16°C) to the disordered polymer (T > 40°C).

Table 1. Dimer formation of TpT in various solvents

Solvent	$\Delta F°$ (kcal/mole)	Formation of TD (%)
t-Butanol	- 2.9	5
Ethanol	- 4.2	5.5
Methanol	- 4.5	6
n-Butanol	- 4.7	7.5
Ethylene glycol	-11.3	19
Formamide	-16.3	21
Glycerol	-16.5	33
Water	-22.3	35

A grating monochromator of Bausch and Lomb, Rochester, was used for the irradiation experiments. Grating: 1200 lines/mm. Light source: Osram HBO 200 W. Wavelength: 280 nm. Ultraviolet dose: 10^5 ergs/mm^2. Concentration of TpT: 25 μg/ml. Room temperature. (Table from ref. 14.)

The production of thymine dimers in DNA is also temperature-dependent (17). In aqueous solutions of native DNA, no change in either base stacking or dimer production occurs until the melting temperature is reached (Fig. 1); at that temperature, transition from the double-stranded to the single-stranded polymer occurs. As shown, a sharp decrease in the yield of dimers occurs upon passing through the melting temperature. The sharpness of this change reflects the cooperative nature of the transition from the stacked double-stranded form to the partially stacked single-stranded form.

On the other hand, the production of dimers in aqueous solutions of single-stranded DNA, which is highly stacked at 25°C in the presence of salt, decreases linearly with increasing temperature. This decrease corresponds to the loss of base stacking, which occurs noncooperatively in the single-stranded polymer.

The presence of ethylene glycol decreases the base stacking of single-stranded DNA and lowers the melting temperature of native DNA. The influence of ethylene glycol on dimer formation in DNA is shown in Fig. 1; the results are consistent with the proposal that base stacking favors dimerization (13, 14).

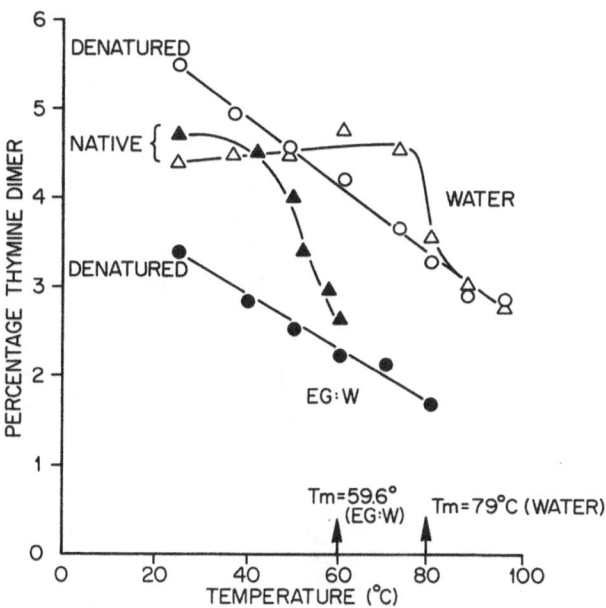

Fig. 1.   Temperature dependence of thymine dimer formation in E. coli DNA at a concentration of 2.5 μg/ml.   Samples labeled "native" were never exposed to temperatures greater than the irradiation temperature.   The incident dose at 254 nm was 1×10^4 ergs/mm^2; the variation of thymine dimer with dose was linear up to 10^4 ergs/mm^2 at both 25°C and 76°C. The melting temperature, T_m, was 20°C lower in a 50% ethylene glycol solution (EG:W) than in water, as determined by separate absorbance measurements at 260 nm. (From ref. 17.)

T < 25°C

When DNA is irradiated at temperatures below that at which the solvent crystalizes, the yield of dimers is reduced (18).   We assume that the rigid environment restricts the motional freedom of the bases and prevents them from assuming the proper orientation for dimerization to occur.   The variation in thymine dimer yield with temperature, going from 25°C to –196°C (19), is shown in Fig. 2 for DNA dissolved in a 50% ethylene glycol solution, which glasses at –130°C (20).   At very low temperatures it appears that only a small fraction of the thymines (< 1%) are favorably oriented for dimerization.   Included in Fig. 2 is the variation in the UV inactivation of transforming DNA with temperature (19).   As shown, the curve for biological survival, which denotes the ability of the irradiated DNA to transform the DNA of cells to which it is added, is the same function of irradiation temperature as the curve for dimer formation.   This similarity supports the claim that thymine dimers are responsible for the UV inactivation of transforming DNA (21).

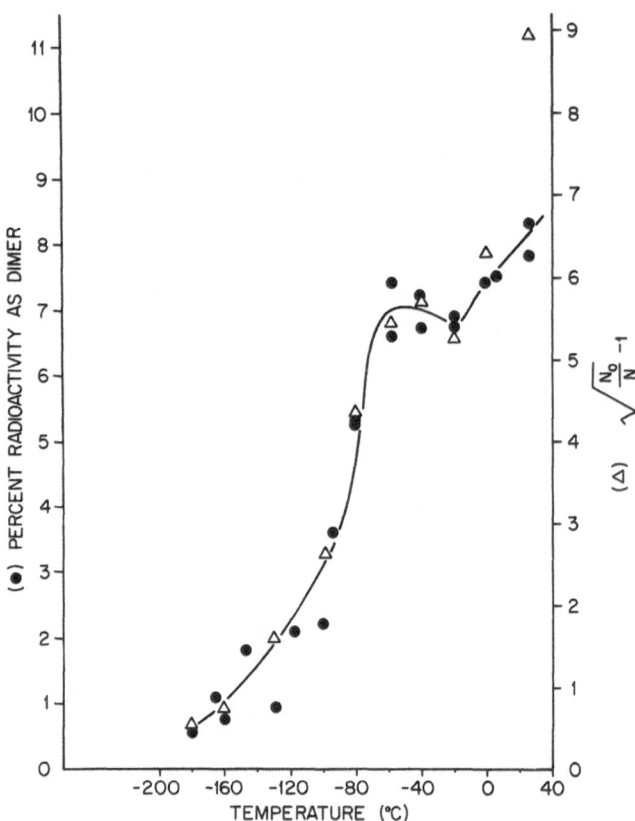

Fig. 2.   Inactivation at 254 nm of H. influenzae transforming DNA
(cathomycin marker) and dimer formation in ^{3}H-labeled H. influenzae DNA
by a dose of 10^4 ergs/mm^2 at different temperatures.   N$_o$ and N are the
number of transformations resulting from the unirradiated and irradiated DNA,
respectively.   The parameter used as a measure of the biological inactivation,
$\sqrt{N_o/N}$ - 1, represents the slope of the UV dose curve at 10^4 ergs/mm^2.
(From ref. 19.)

On the other hand, short wavelength irradiation (λ < 250 nm) at -196°C
of either DNA or poly dT containing a large concentration of dimers leads to
no appreciable reversal of the dimers (22).   That is, a dimer once broken at
-196°C is reformed very easily at that temperature because the rigid environ-
keeps the two residues favorably oriented for redimerization.   Because of
this failure of broken dimers to move apart in a rigid solvent, it is possible at
low temperatures to increase the yield of dimers over that normally obtained
at 25°C with short wavelengths of irradiation.   If one thaws and refreezes a
sample containing dimers, then reirradiation will establish new dimers and not

destroy the old ones.   As shown in Fig. 3, irradiation of poly dT at -196°C, using a freezing and thawing cycle, leads to twice as many dimers as that normally obtained at 25°C for this wavelength of irradiation (240 nm).

Another photoproduct, chromatographically similar to that isolated from irradiated bacterial spores, is obtained when isolated DNA is irradiated at low temperatures (18, 19).   This spore-type photoproduct, whose yield is a maximum at ~ -100°C (Fig. 4), is not affected by treatment with the photoreactivating enzyme (PR) — this treatment monomerizes pyrimidine dimers.   PR of DNA which has been irradiated at various temperatures leads to a residual level of inactivation, as shown in Fig. 4, even though all dimers are removed.   The irradiation temperature at which maximal inactivation occurs following PR is ~ -100°C, the same temperature at which the spore photoproduct formation reaches a maximum.   We conclude from these results that the spore photoproduct makes a significant contribution to the inactivation of transforming DNA irradiated at low temperatures.

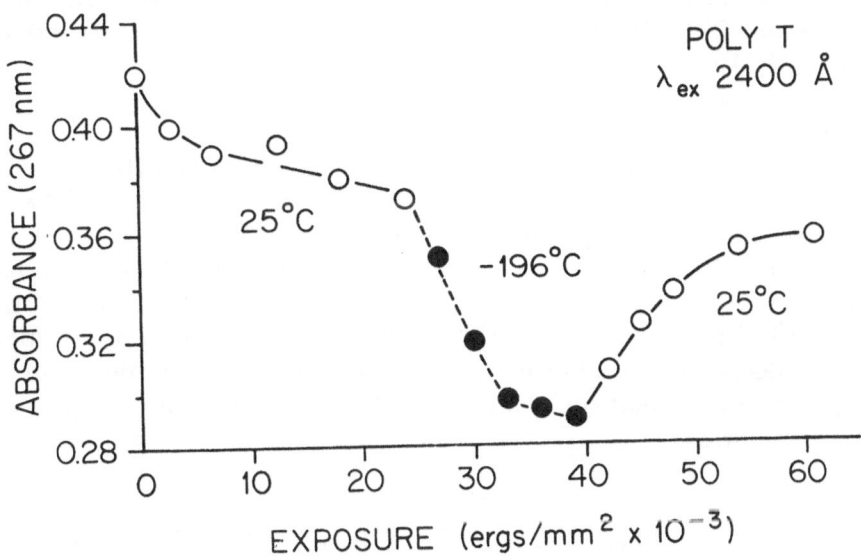

Fig. 3.   Absorbance changes (25°C) upon irradiation (240 nm) of poly dT in a 50% ethylene glycol solution.   Following the initial irradiation at 25°C, the sample was irradiated at -196°C and thawed to 25°C and the absorbance was measured.   This procedure was continued until a low-temperature steady-state concentration of dimer was reached.   Finally, the sample was reirradiated at 25°C. ·

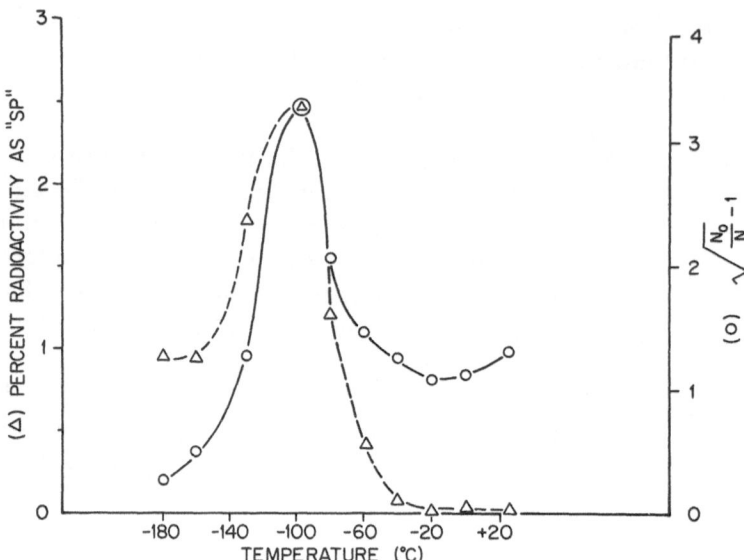

Fig. 4. Comparison of "sp" formation at 254 nm following a dose of 10⁴ ergs/mm² at different temperatures with the residual inactivation of H. influ-enzae transforming DNA (streptomycin marker) after maximum photoreactiva-tion. (From ref. 19.)

The temperature dependence of photoproduct formation and UV survival of bacterial spores of Bacillus megaterium was also measured (23), and the re-sults are presented in Fig. 5. Only spore-photoproduct formation was observed and, as in isolated DNA, it shows a maximum yield at -100°C, the same tem-perature at which the UV sensitivity of the spores is a maximum. Below this temperature both the photoproduct yield and the UV inactivation of the spores decreases with decreasing irradiation temperature. These temperature-dependence studies help demonstrate that the UV inactivation of spores is due to spore-photoproduct formation.

RELATIVE HUMIDITY (r.h.)

The conformation of DNA depends upon the amount of water associated with it; as the amount of bound water decreases, DNA goes from the B confor-mation to a disordered form (24). In order to determine how such a transition affects the photochemistry of DNA, films of DNA were irradiated at various relative humidities between 0 and 100%, and the concentration of thymine photoproducts was measured (25). At high r.h., DNA is in the B form and its

photochemistry is the same as that in solution (Fig. 6).   A cooperative transition in the photoproduct yields occurs when the transition from the B form to
the disordered form takes place; i.e., below 65% r.h.   In the disordered form,
thymine dimer production is reduced while spore-photoproduct production is
considerably enhanced.   It is possible that the DNA in bacterial spores is also
in such a "dry" and disorganized state.   The two different curves obtained for
each photoproduct depend upon whether the equilibration of hydration was
reached from high or low r.h. and reflect the absorption-desorption hysteresis
observed for water binding (24).

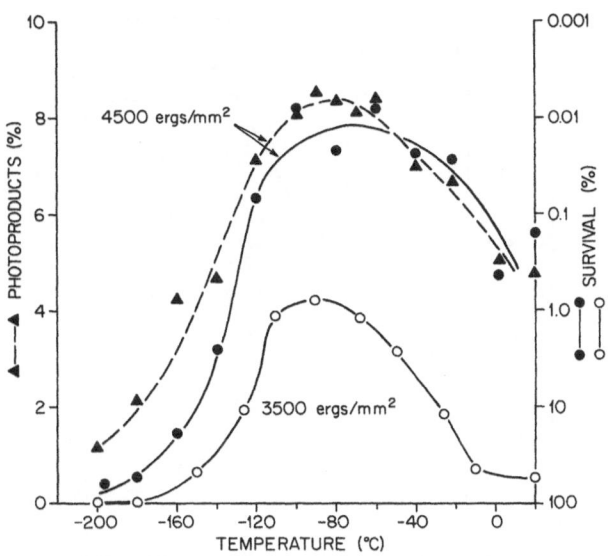

Fig. 5.   Temperature dependence of the survival (O, ●) and the amount
of thymine-containing photoproducts (b and c) ▲ in B. megaterium spores
exposed to fixed doses (4,500 ergs/mm^2, ●, ▲ ; 3,500 ergs/mm^2, O) of 254-
nm irradiation.   Note that percent survival increases from top to bottom on
right-hand ordinate.   (From ref. 23.)

pH

In alkaline solution (pH 11-12) thymidine is ionized at the N_3 position
and, as Sztumpf and Shugar (26) showed, irradiation of ionized TpT results in
a shift of the photoinduced steady-state equilibrium from dimer to monomer.
We have examined the influence of high pH on dimer formation in poly dT.
As shown in Fig. 7a, the rate of dimerization at pH 13 is considerably less

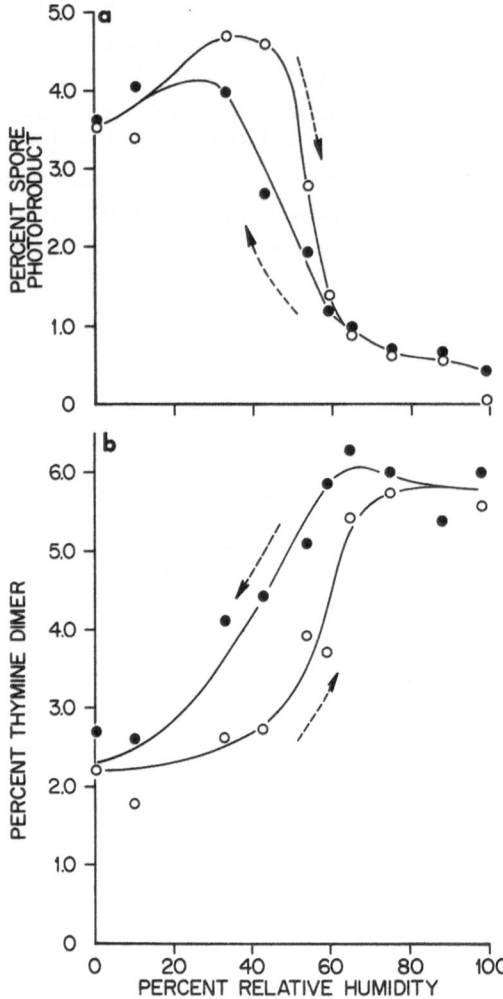

Fig. 6. Photoproduct formation in DNA films equilibrated at various relative humidities. Films were irradiated at 254 nm with 60,000 ergs/mm^2. The data on the adsorption curve represents an average of two different experiments. The arrows indicate the path by which the final relative humidity was obtained.

than it is at pH 7. Part of the reduction in dimer yield at pH 13 may be due to a higher rate of dimer reversal because of enhancement of the dimer absorbance at long wavelengths for this pH (27). However, it appears from the difference in the initial rate of dimerization that the forward reaction is also impaired, possibly because of the negative charges on each base at this pH. These charges would set up repulsive forces at the distance required for dimerization.

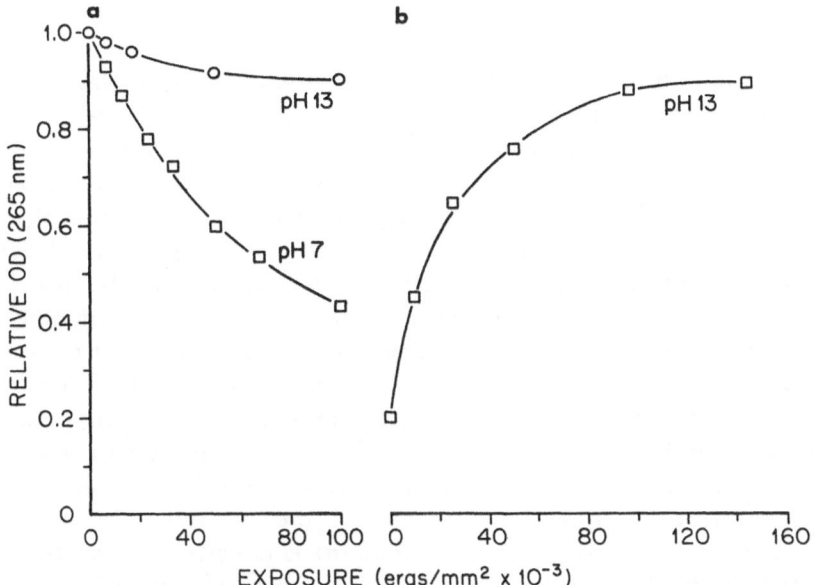

Fig. 7. (a) Absorbance changes following irradiation (280 nm) of poly dT either at pH 7 or at pH 13. (b) Irradiation (280 nm) of poly dT previously irradiated (280 nm) at pH 7.

Irradiation at pH 13 of a sample previously irradiated at pH 7 results in the reversal of nearly all of the dimers (Fig. 7b); the final steady-state concentration of dimers at pH 13 is the same regardless of whether the sample was previously irradiated at pH 7 or not.

EFFECTS OF BOUND MOLECULES ON THE PHOTOCHEMISTRY OF DNA

Dyes

The organic dyes acridine orange and proflavin bind to DNA and reduce the formation of thymine dimers by about an order of magnitude (28, 29). As a result, thymine dimers are monomerized when previously irradiated DNA is reirradiated in the presence of a bound dye. A mechanism has been proposed by Sutherland and Sutherland (30) for the inhibition by dyes of thymine dimerization. It involves the transfer of singlet energy from thymine to the bound dye; i.e., the dye is an energy sink that reduces the concentration of thymine in the excited-singlet state and leads to lower dimer yields. The dye has no effect on the rate of monomerization of the dimers with UV (29).

Mercuric Ions

Mercuric ions belong to a class of metal ions which bind to DNA at a site located on the heterocyclic bases and not on the phosphate diester. The binding sites on the bases are such that protons involved normally in hydrogen bonding are displaced and the bound metal ion interacts with the π electron cloud of the base. Consequently, the physical and spectral properties of DNA are drastically altered when mercuric ions are bound (31). The effect of bound mercuric ions on the photochemistry of DNA has recently been investigated (32). As shown in Fig. 8a, the thymine dimer yield is reduced when DNA is irradiated with mercuric ions present. The addition of excess chloride reverses the mercuric ion binding and eliminates the inhibition of dimerization. Since bound mercuric ions do not appreciably affect the rate of thymine dimerization in both TpT and poly dT (32), it appears that in DNA the binding of mercuric ions to bases other than thymine, such as adenine or guanine, makes potential energy traps which quench the excited states of thymine and lead to lower dimer yields. This proposal is consistent with the observation (33) that Hg(II) quenches the phosphorescence of thymine in DNA when it is bound to bases other than thymine. Hg(II) also quenches the phosphorescence of poly G and poly A, but no such quenching occurs in poly dT.

As indicated in Fig. 8b, dimer reversal occurs when DNA that contains thymine dimers is irradiated with mercuric ions present. This effect should prove useful when it is desirable to photoreverse the amount of dimer present in DNA without changing the irradiation wavelength. Experiments are currently being done with irradiated transforming DNA to test whether reversing dimers in this fashion leads to an increase in the ability of the DNA to transform.*

SUMMARY

The photochemistry of DNA and poly dT has been examined under a variety of experimental conditions. In particular, the formation of the thymine dimer and the spore photoproduct was measured and found to be strongly dependent upon the solvent, temperature, relative humidity, pH, and presence of bound molecules during irradiation. Often the variation in photochemical response seems directly related to some change in the conformation of the DNA, which suggests the use of photochemical response as a tool for structural investigations of macromolecules. Finally, irradiation of previously irradiated DNA, either in the presence of proflavin or Hg(II), or at high pH, results in considerable reversal of the dimers initially present.

---

*Note added in proof: We have found (32) that reversal of the UV-induced biological inactivation of transforming DNA occurs when mercuric ions are added and the DNA is irradiated further.

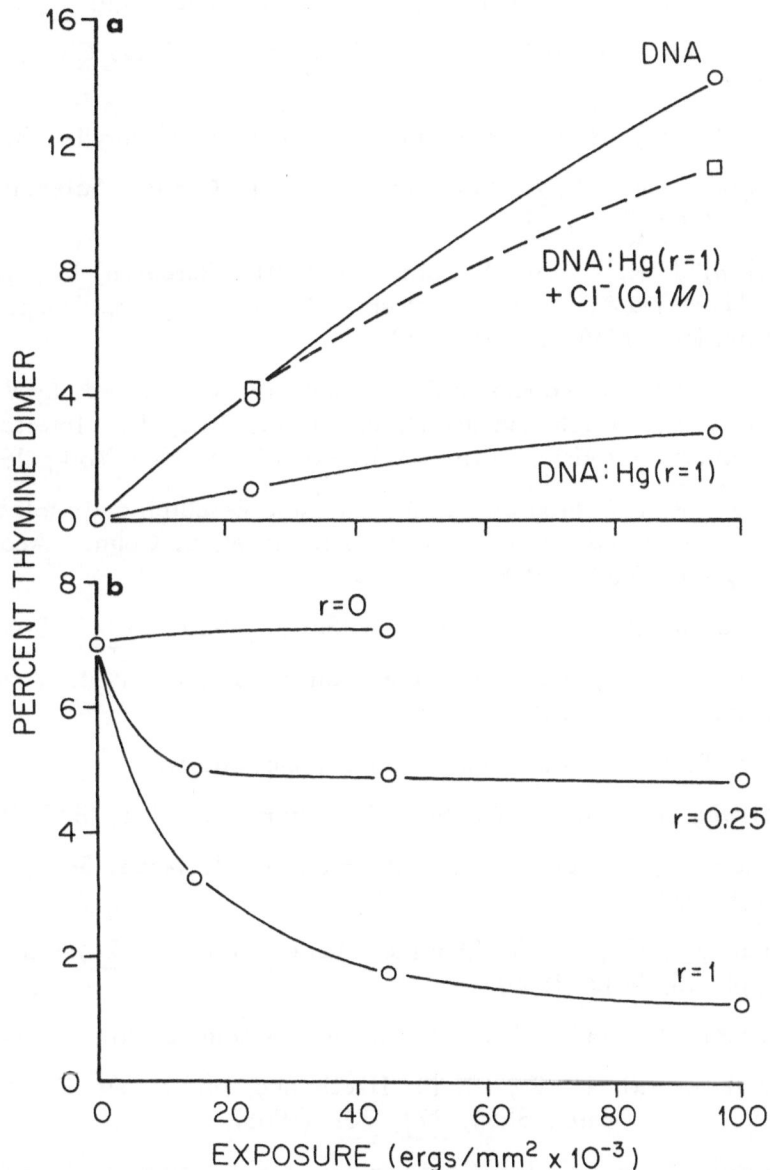

Fig. 8. (a) Influence of $HgCl_2$ on production of thymine dimers in E. coli DNA. $r = [Hg]/[PO_4]_{DNA}$; $\lambda_{ex} = 280$ nm. (b) Photoreversal of thymine dimers in DNA upon addition of $HgCl_2$. $\lambda_{ex} = 254$ nm.

REFERENCES

1. Beukers, R., and W. Berends, Biochim. Biophys. Acta $\underline{41}$, 550 (1960).

2. Beukers, R., J. Ijlstra, and W. Berends, Rec. Trav. Chim. $\underline{79}$, 101 (1960).

3. Wacker, A., H. Dellweg, and D. Weinblum, Naturwiss. $\underline{47}$, 477 (1960).

4. Setlow, R. B., P. A. Swenson, and W. L. Carrier, Science $\underline{142}$, 1464 (1963).

5. Setlow, J. K., Current Topics in Radiation Research, Vol. II, ed. by M. Ebert and A. Howard. North-Holland Publishing Company, Amsterdam, 1966, pp. 195-248.

6. Burr, J. G., Advances in Photochemistry, Vol. 6, ed. by W. A. Noyes, Jr., G. S. Hammond, and J. N. Pitts, Jr. Interscience Publishers, a division of John Wiley and Sons, New York, 1968.

7. Setlow, R. B., Progress in Nucleic Acid Research and Molecular Biology, Vol. 8, ed. by J. N. Davidson and W. E. Cohn. Academic Press, New York, 1968.

8. Donnellan, J. E., Jr., and R. B. Setlow, Science $\underline{149}$, 308 (1965).

9. Stafford, R. S., and J. E. Donnellan, Jr., Proc. Natl. Acad. Sci. U.S. $\underline{59}$, 822 (1968).

10. Rahn, R. O., and L. Landry, unpublished results.

11. Dellweg, H., and A. Wacker, Z. Naturforsch. $\underline{17b}$, 827 (1962).

12. Lamola, A. A., and T. Yamane, Proc. Natl. Acad. Sci. U.S. $\underline{58}$, 443 (1967).

13. Sinanoglu, O., and S. Abdulnur, Federation Proc. $\underline{24}$, No. 2, Part III, Suppl. 15, S-12 (1965).

14. Wacker, A., and E. Lodeman, Angew. Chem. Intern. Ed. $\underline{4}$, 150 (1965).

15. Zavil'gel'skii, G. B., B. N. Ilyashenko, and S. Ya. Dityatkin, Dok. Akad. Nauk, SSSR, $\underline{171}$, 732 (1966).

16. Tramer, Zofia, K. L. Wierzchowski, and D. Shugar, Acta Biochim. Polon. $\underline{16}$, 83 (1969).

17. Hosszu, J. L., and R. O. Rahn, Biochem. Biophys. Res. Commun. $\underline{29}$, 327 (1967).

18. Rahn, R. O., and J. L. Hosszu, Photochem. Photobiol. $\underline{8}$, 53 (1968).

19. Rahn, R. O., J. K. Setlow, and J. L. Hosszu, Biophys. J. $\underline{9}$, 510 (1969).

20. Luyet, B., and D. Rasmussen, Biodynamica 10, 167 (1968).

21. Setlow, R. B., and J. K. Setlow, Proc. Natl. Acad. Sci. U.S. 48, 1250 (1962).

22. Rahn, R. O., and J. L. Hosszu, Photochem. Photobiol. 7, 637 (1968).

23. Donnellan, J. E., Jr., J. L. Hosszu, R. O. Rahn, and R. S. Stafford, Nature 219, 964 (1968).

24. Falk, M., K. A. Hartman, Jr., and R. C. Lord, J. Am. Chem. Soc. 85, 391 (1963).

25. Rahn, R. O., and J. L. Hosszu, Biochim. Biophys. Acta 190, 126 (1969).

26. Sztumpf, E., and D. Shugar, Biochim. Biophys. Acta 61, 555 (1962).

27. Setlow, R. B., Science 153, 379 (1966).

28. Beukers, R., Photochem. Photobiol. 4, 935 (1965).

29. Setlow, R. B., and W. L. Carrier, Nature 213, 906 (1967).

30. Sutherland, B. M., and J. C. Sutherland, Biophys. J. 9, 292 (1969).

31. Yamane, T., and N. Davidson, J. Am. Chem. Soc. 83, 2599 (1969).

32. Rahn, R. O., and L. Landry, Biochim. Biophys. Acta, submitted for publication.

33. Rahn, R. O., M. Battista, and L. Landry, Biochim. Biophys. Acta, submitted for publication.

THE PHOTOCHEMICAL ADDITION OF AMINO ACIDS AND PROTEINS TO NUCLEIC ACIDS

Kendric C. Smith

Department of Radiology, Stanford University School of

Medicine, Stanford, California 94305

ABSTRACT: When bacterial cells are irradiated with ultraviolet light, DNA and protein become photochemically cross-linked. This photochemical reaction appears to be the major lethal lesion in irradiated bacteria under certain experimental conditions. We have turned to in vitro model systems in order to gain some insight into the mechanism(s) by which DNA and protein are photochemically cross-linked in vivo. A survey was performed of the ability of the 22 common amino acids to add photochemically (254 nm) to ^{14}C-uracil. The 11 reactive amino acids were glycine, serine, phenylalanine, tyrosine, tryptophan, cystine, cysteine, methionine, histidine, arginine and lysine. The most reactive amino acids were phenylalanine, tyrosine and cysteine. We have isolated the cysteine adduct and have shown it to be 5-S-cysteine-6-hydrouracil. The analogous thymine adduct has also been isolated and its tentative structure is 5-S-cysteine-6-hydrothymine. We have studied the kinetics of the photochemical addition of ^{35}S-cysteine to various synthetic and natural polynucleotides. Preliminary data on the tyrosine adducts to uracil are reported.

INTRODUCTION

When bacterial cells are irradiated with ultraviolet light, there is a dose-dependent decrease in the amount of deoxyribonucleic acid (DNA) that can be isolated in pure form from these irradiated bacteria (1, 2). This phenomenon arises as a consequence of the photochemical cross-linking of the DNA with protein. This cross-linking of DNA and protein appears to be the major lethal lesion in irradiated bacteria under certain experimental conditions.

31

For example, the enhanced yield of DNA cross-linked to protein for
a given dose of ultraviolet radiation may explain the enhanced
sensitivity of bacteria to killing by UV under conditions of thymine
starvation (3) or when cells are irradiated while frozen (4). The
chemical mechanism by which DNA and protein are photochemically
cross-linked in vivo is not known. Because of the complexities of
dealing with cellular systems, we have turned to in vitro model
systems in order to study the possible mechanisms by which DNA and
protein may become photochemically cross-linked.

EXPERIMENTAL

The two photochemical reactions of the pyrimidines that are
best understood are the formation of cyclobutane-type dimers
through carbon atoms 5 and 6 and the hydration reaction which
involves the addition of a molecule of water across the 5-6 double
bond of single pyrimidines (for recent reviews see 5-7). We therefore wondered whether reactions of these types might be involved
in the photochemical cross-linking of DNA and protein. We have
tried to form cross-adducts between thymine and the several aromatic
amino acids by irradiating suitable mixtures in frozen solution;
conditions which favor the dimerization of thymine when irradiated
by itself. These attempts have thus far proved unsuccessful. We
have also irradiated various OH and SH amino acids in solution
with uracil in the hope of finding reactions analogous to the
addition of water across the 5-6 double bond. Early experiments
indicated that the SH amino acid, cysteine, was very reactive in
photochemically combining with uracil. We have isolated this
photoproduct and determined its structure to be 5-S-cysteine-6-
hydrouracil (Figure 1).

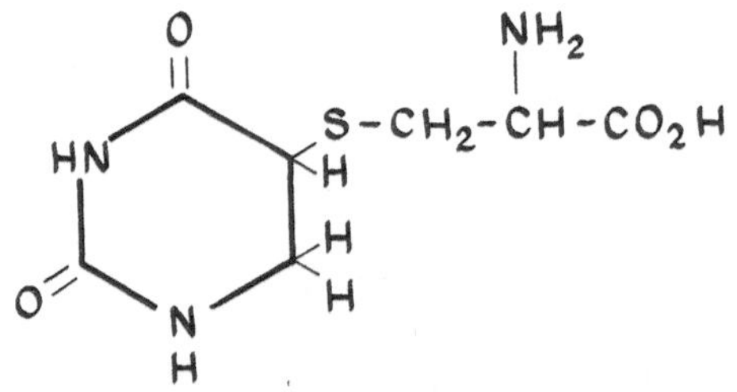

Figure 1. 5-S-cysteine-6-hydrouracil (8).

Ultraviolet, infrared, nuclear magnetic resonance, and mass
spectroscopy were used to identify this photoproduct (8). Treating
the photoproduct with deuterated Raney nickel yields 5-mono-
deuteriodihydrouracil, thus confirming the point of attachment of
the cysteine. Raney nickel treatment of the cross-adduct also
yields alanine; the desulfurated product of cysteine. 5-S-
cysteine-6-hydrouracil-HCl is stable to heat (100°C) in water

solution and is stable to 6 N HCl at room temperature but is not stable to the heat and acid conditions used for the hydrolysis of DNA. It is quite unstable to alkali.

To investigate the scope of the raction of cysteine with the various pyrimidines we have determined the rate constants for the uptake of cysteine by various synthetic and natural polynucleotides per dose of ultraviolet radiation (9). These results are summarized in Table 1.

TABLE I: Rate Constants for the Photochemical Addition of [^{35}S]Cysteine to Polynucleotides.

Polynucleotide	K^a	
	Exptl	Calcd
Poly rU	21.8 (13.3)^d	
Poly rU:rA	0.7^d (U only)	
Poly rA	0.6	
Poly rC	8.1 (0.6)^d	
RNA (yeast)^b	21.8^e	
	4.8^e	
Poly dC	2.6	
Poly dC:dG	2.6 (C only)	
Poly dT	5.4	
Poly dT:dA (heated)^c	4.2 (T only)	
Poly dT:dA	2.6 (T only)	
Poly dAT:dAT	1.1 (T only)	
DNA (calf thymus)	2.6^f	2.6^g
DNA (heated)^c	4.2^f	4.2^h

^{a} K = (μmoles of cysteine/μmole of PO$_4$ involved)/ ergs/mm^2 × 10^8 (at pH 5). ^{b} RNA shows a biphasic uptake of cysteine. ^{c} For 15 min at 100° in 0.075 M NaCl. Quick cooled. ^{d} At pH 6.5. ^{e} For (20% C + 27% U). ^{f} For (21% C + 29% T). ^{g} For (dC:dG + dA:dT). ^{h} For (dC + dT). (9).

Cysteine readily combines with polymers containing uridylic acid, cytidylic acid or thymidylic acid. Whether the polymers are single or double-stranded (and/or protonated) has a profound effect upon the reaction rate.

Whereas the cysteine that was photochemically combined with polyuridylic acid was completely stable to heating at 65°C for 60 minutes, there was a 16% loss of cysteine from polydeoxythymidylic acid and a 49% loss from polydeoxycytidylic acid. Using these percentage lability figures and correcting for the thymine and cytosine content of the DNA, we calculated that 30% of the cysteine photochemically attached to calf thymus DNA should be heat labile. This was in good agreement with the value of 36% observed experimentally. These results suggest that there are at least two mechanisms for the photochemical linkage of cysteine to cytidylic acid and to thymidylic acid in these polymers. The heat lability of certain of the amino acid adducts to DNA may help to explain some of our problems in dealing with the whole bacterial system where we have heated the cells in detergent at about 65°C in order to lyse the cells and to denature their proteins.

We have isolated the cysteine adduct to thymine, and based upon nuclear magnetic resonance spectroscopy, the structure of this photoproduct appears to be analagous to the uracil product and has been identified as 5-S-cysteine-6-hydrothymine (10).

We have observed that dihydrouracil is formed in significant quantities when irradiated in the presence of SH compounds such as cysteine, cysteamine and H_2S (8). Recently, Jellinek and Johns (11) have undertaken a study of the chemical mechanisms involved in the photochemical addition of cysteine to uracil and in the formation of dihydrouracil. Their data suggest that the triplet excited state of uracil can abstract hydrogen atoms from cysteine to form dihydrouracil. The thiyl radicals generated by this process can add to ground state uracil molecules to yield the cross-adduct between cysteine and uracil.

Two reactions of the pyrimidines that have been described by other authors may have some importance in the photochemical cross-linking of DNA and protein. Alcantara and Wang (12) have observed the formation of 5-formyluracil when thymine in aqueous solution is exposed to ultraviolet radiation. If this reaction occurs in irradiated DNA, the formyl group could then react with a protein amino group to give a covalent bond between DNA and protein (13).

Janion and Shugar (14) have observed that dihydrocytosine will react with glycine such that the amino group of the dihydrocytosine is replaced by the amino group of the glycine, resulting in a covalent link between dihydrocytosine and glycine. Since both the photohydrate and cyclobutane-type dimer of cytosine are analogs of dihydrocytosine, one may predict that the addition of protein amino groups to these cytosine photoproducts might serve as another mechanism by which DNA and protein are cross-linked by ultraviolet radiation.

Several lines of evidence have suggested that cysteine is probably not the only amino acid capable of combining photochemically with the nucleic acids. Our first line of evidence was the observation that the protein gelatin, which contains no cysteine, does photochemically cross-link with DNA in vitro, albeit at a much reduced efficiency compared with bovine serum albumin which does contain cyst(e)ine (15). In order to determine the scope of the reactivity of amino acids with the nucleic acids we have investigated the ability of the 22 common amino acids to add photochemically (254 nm) to ^{14}C-uracil (16). The 11 reactive amino acids were glycine, serine, phenylalanine, tyrosine, tryptophan, cystine, cysteine, methionine, histidine, arginine and lysine (Figure 2).

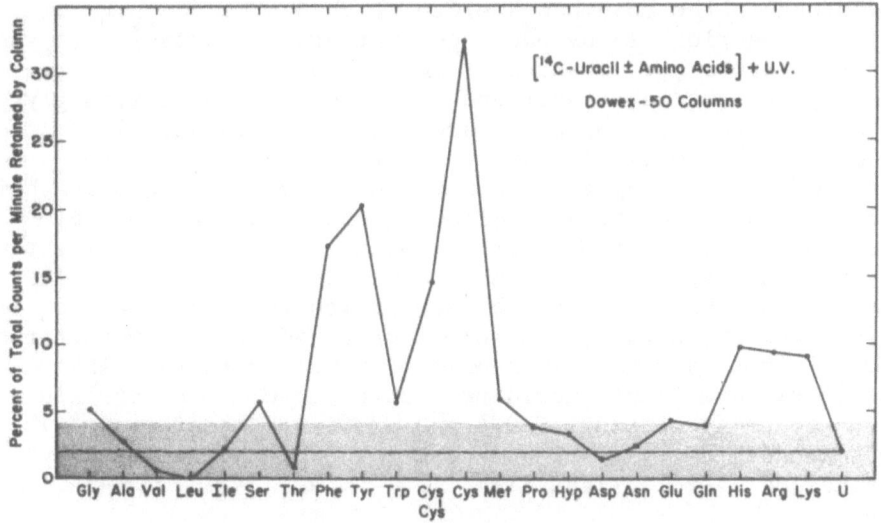

Figure 2. The Photochemical Addition of Amino Acids to ^{14}C-Uracil. A 0.2 ml aliquot of amino acid solution (0.01 M; except tyrosine at 0.003 M) was mixed with 0.05 ml of ^{14}C-2-uracil (0.0011 M; 25 µCi/ml). The molar ratio of amino acid to uracil was thus ≈36:1 (except for tyrosine at ≈11:1). The solution was irradiated for 200 minutes in a Pyrex spot plate in contact with the rim of a Mineralight U.V. lamp (Model UVS-11, Ultra-Violet Products, Inc.) whose output is mainly at 254 nm. An aliquot was then assayed for content of radioactivity (liquid scintillation counter) and 0.05 ml was introduced to a 4 ml column of Dowex-50-. HCl in a plastic 5 ml syringe. The column was rinsed with 25-one ml portions of water (into a volumetric flask) and an aliquot of the combined effluent was counted for radioactivity. The 100% sample minus the material that did not stick to the column gives the amount retained by the column. Most of these counts could be recovered by eluting with 6 N HCl. The results plotted here are

the average of 2-3 experiments (and 9 for ^{14}C-uracil (U) alone).
The hatched area indicates the spread of the data for ^{14}C-uracil
irradiated in the absence of amino acids. (16)

The most reactive amino acids were phenylalanine, tyrosine and
cysteine. The finding that 11 amino acids react photochemically
with uracil suggests many new mechanisms by which DNA and protein
can become cross-linked in vivo by ultraviolet radiation.

Having elucidated the photochemical reaction between cysteine
and uracil we are currently investigating the nature of the
photochemical reaction between tyrosine and uracil. A mixture of
tyrosine and ^{14}C-uracil was irradiated (254 nm) and then poured
through an ion exchange column that reacts with basic groups
(Dowex 50). The column was purged with water and then the material
that had stuck to the column was eluted with 6 N HCl. When this
material was fractionated by paper chromatography, approximately
6 radioactive spots were observed. Several of these spots were
UV-absorbing and one major spot and one minor spot were strongly
ninhydrin positive. One major product has been tentatively
identified as dihydrouracil. We are currently attempting to
isolate a sufficient amount of each of these products so that their
chemical structures may be determined. Concerning the possible
mechanisms of these reactions with tyrosine, it is interesting that
the presence of oxygen greatly facilitates these reactions. It
therefore may be that some type of sensitized reaction that requires
oxygen is involved in the formation of cross-adducts between tyrosine
and uracil. In this regard it should be recalled that in the
presence of visible light, acridine orange (1) or methylene blue
(5) cause the cross-linking of DNA and protein in bacteria. Also,
the cross-linking of cysteine and uracil has been sensitized by
riboflavine and visible light (11).

                              CONCLUSIONS

In the past, the photochemistry of the nucleic acids and of
the proteins have been studied separately. However, in living
cells, the nucleic acids and the proteins are not present in
separate compartments but are in constant interaction. It thus
seems logical that if there are photochemical reactions between
proteins and nucleic acids that these interactions may have significant biological consequences in cells that have been irradiated
with ultraviolet light. This appears to be the case.

A new class of compounds have been described; cross-adducts
of amino acids and pyrimidines. Since about half of the common
amino acids readily cross-link photochemically with uracil, the
generality of this type of reaction is established. The mechanisms
for the addition of these amino acids to uracil must await the

bulk isolation of the products and their chemical identification.

The observation that the nucleic acids can combine photo-chemically with amino acids suggests that more attention should be given to these addition reactions in attempts to explain the biological effects of ultraviolet radiation, especially when the effects cannot be adequately explained by the known photoproducts produced in pure DNA.

REFERENCES

1. K. C. Smith, Biochem. Biophys. Res. Commun. 8, 157 (1962).
2. K. C. Smith, Photochem. Photobiol. 3, 415 (1964).
3. K. C. Smith, B. Hodgkins and M. E. O'Leary, Biochim. Biophys. Acta 114, 1 (1966).
4. K. C. Smith and M. E. O'Leary, Science 155, 1024 (1967).
5. K. C. Smith and P. C. Hanawalt, Molecular Photobiology: Inactivation and Recovery, Academic Press, New York (1969).
6. K. C. Smith, Radiation Res. Suppl. 6, 54 (1966).
7. R. B. Setlow, Prog. Nucleic Acid Res. Mol. Biol 8, 257 (1968).
8. K. C. Smith and R. T. Aplin, Biochemistry 5, 2125 (1966).
9. K. C. Smith and D. H. C. Meun, Biochemistry 7, 1033 (1968).
10. K. C. Smith, Manuscript in preparation.
11. T. Jellinek and R. B. Johns, Photochem. Photobiol., In press.
12. R. Alcantara and S. Y. Wang, Photochem. Photobiol. 4, 473 (1965).
13. S. Y. Wang and R. Alcantara, Photochem. Photobiol. 4, 477 (1965).
14. C. Janion and D. Shugar, Acta Biochim. Polon. 14, 293 (1967).
15. K. C. Smith, in Radiation Research (G. Silini, ed.) North Holland Publ. Co., Amsterdam (1967), p. 756.
16. K. C. Smith, Biochem. Biophys. Res. Commun. 34, 354 (1969).

ENERGY TRANSFER IN POLYMERIC KETONES IN THE SOLID PHASE

M. Heskins and J.E. Guillet

Department of Chemistry
University of Toronto
Toronto 181, Canada

Energy transfer studies in polymeric systems have indicated that effective stabilization of polymer molecules against the action of ultra-violet radiation may be possible by this mechanism. [1, 2, 3] However, most of the definitive studies on the quenching of excited states in polymers have been carried out in dilute solution, and it seemed important to establish that similar energy exchange is possible in the bulk polymer phase. Studies were therefore carried out on the exchange of electronic energy to and from the ketone carbonyl chromophore in solid polymer films. The polymers used were copolymers of ethylene and carbon monoxide containing about 1% ketone carbonyl in the backbone of what is essentially a branched polyethylene chain. The photochemistry of these polymers has been studied extensively by Hartley and Guillet [4, 5] and it was shown that two major photochemical reactions occurred, the Type I giving two free radicals, and the Type II producing a double bond and lower ketone. In solution both reactions cause a reduction in molecular weight of the polymer. Energy exchange studies in hydrocarbon solution were reported more recently by Heskins and Guillet. [1, 2]

EXPERIMENTAL

The materials and apparatus used were the same as reported by Heskins and Guillet. [1, 2] Molecular weights were estimated viscometrically, and product analysis carried out by infra-red or gas-chromatographic methods where applicable.

Film Measurements

Films were compression-molded from the hexane-extracted beads in a Carver Press for 1 minute at 15,000 psi. The films were rapidly quenched by passing cold water through the press. Various concentrations of COD were incorporated into the film by soaking the film in COD for 24 hours at different temperatures. The films were then wiped free of surface additive and clamped between two quartz plates and placed in a special cell. The cell was pressurized to 50 psi Nitrogen to ensure minimum loss of COD from the films. The film was then irradiated for a convenient length of time after which the film was extracted with pentane to remove the COD which was determined spectrophotometrically to estimate the concentration of COD originally present in the film. The concentration of vinyl groups in the extracted film was determined by the method of Hartley and Guillet[1] after drying under vacuum at room temperature. The above sequence was repeated 3-4 times for each film using the same area for irradiation in successive measurements.

Triplet Quenching of Polymer Photolysis

1.3 cyclo octadiene (COD) was used for triplet quenching studies. It absorbs very weakly at 313 nm ($\varepsilon_{313} = 0.03$) and this is probably singlet-triplet absorption. Its triplet level is not known but has been placed at approximately 70 kcal/mole[6] based on the fact that it accepts triplet energy from acetophenone. Since the triplet level of aliphatic ketones is expected to be greater than that of acetophenone, energy transfer between the polymeric ketone and COD should also be efficient. However it is unlikely that COD will quench excited singlet states of ketones because the energy level of its first excited singlet lies well above that of the ketone chromophore.

In experimental studies of electronic energy exchange it is convenient to express the experimental results in the form of a Stern-Volmer equation of the form:

$$\frac{\emptyset_0}{\emptyset} = 1 + k_q \, \tau \, [Q]$$

where $\emptyset_0$ is the quantum yield for a particular process in the absence of a quenching molecule

$\emptyset$ is the quantum yield of the quenched process

k_q is the bimolecular rate constant for the quenching process

τ is the lifetime of the state in the absence of quencher

and [Q] is the concentration of the quenching molecule.

Data on the quenching of the photolysis of the ethylene-carbon monoxide

copolymer by COD in solution have been reported in a previous communication.[1]   The Stern-Volmer plot is curved and reaches a limiting
value which indicates that under the conditions of the experiment, only
45% of the total reaction arises from the triplet state, and can be
quenched by COD.   When allowance is made for this, a more linear
Stern-Volmer plot is obtained.   From the limiting slope of this curve
a value of $k_q \tau$ = 20 1 mole^{-1} is derived.

  Similar data were obtained by quenching studies in the polymer
film.   Due to the fact that absorption of COD masks the vinyl group
absorption in the infrared region, the type II quantum yield for the
polymers could not be measured without first removing the COD.
Because very little quenching was observed, the process of alternately
adding and removing COD led to large scatter in the data.   Values of
ΔK, the increase in terminal vinyl absorption are plotted against
light absorbed for the 1% carbonyl film containing two different concentrations of COD and are shown in Fig. 1.   A least squares fit of
the data to a straight line gives the quantum yields shown in Table I.

<u>Figure 1</u>

Quenching of $\emptyset_{II}$ in Polymer Films at 80° C

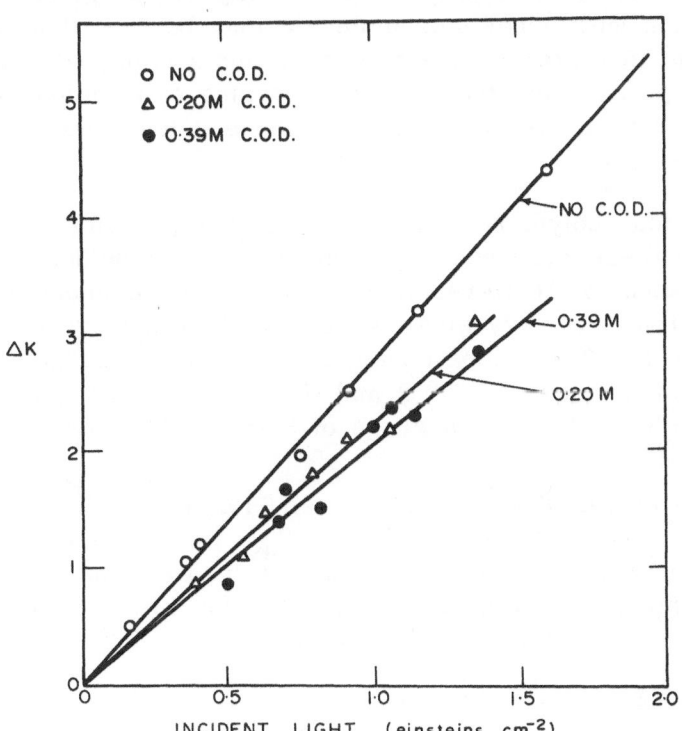

TABLE I

Quenching of Type II Quantum Yield in Polymer Film at 80°C.

Concentration COD mole l^{-1}	ϕ_{II}
0.0	0.033
0.107	0.028
0.20	0.026
0.39	0.025
0.45	0.021

From these data, and assuming that 45% of the reaction comes from the triplet (as in solution), a Stern-Volmer plot yields a value of $k_q\tau = 3$ mole^{-1}.

Fluorescence Quenching by Polymeric Ketones

In order to determine the lifetime of the ketone triplet, it is necessary to determine the rate constant of energy exchange k_q. Since k_q is often considered to be diffusion controlled[7] it could be calculated from the diffusion coefficients for the reacting species. However in view of the fact that the probability of energy transfer p during a collision with a polymer molecule may be less than unity, it was desirable to make an independent estimate from data on the rather similar process involved in the quenching of naphthalene fluorescence, where a direct measure of the efficiency could be obtained.

Films of four polymers containing 0%, 0.3%, 0.55% and 1% carbonyl groups were prepared to contain 0.6M naphthalene. There was a large amount of scattered light at 45° to the incident beam, but at 350 nm this was very small while the fluorescence of naphthalene could be observed. The fluorescence quenching by these polymers was measured at 350 nm and 80°C and the Stern-Volmer plot is shown in Fig. 2. Each point is the average of 3 determinations.

From the plot $k_q\tau = 12$ l mole^{-1} and for $\tau = 8.5 \times 10^{-8}$ sec this gives 1.4×10^8 l mole^{-1} sec^{-1} for k_q, the rate constant for energy transfer, a value about 1/20 that obtained for the polymer in solution.[2] This lower value is undoubtedly due to the reduced

Figure 2. Stern-Volmer Plot for Quenching of Naphthalene

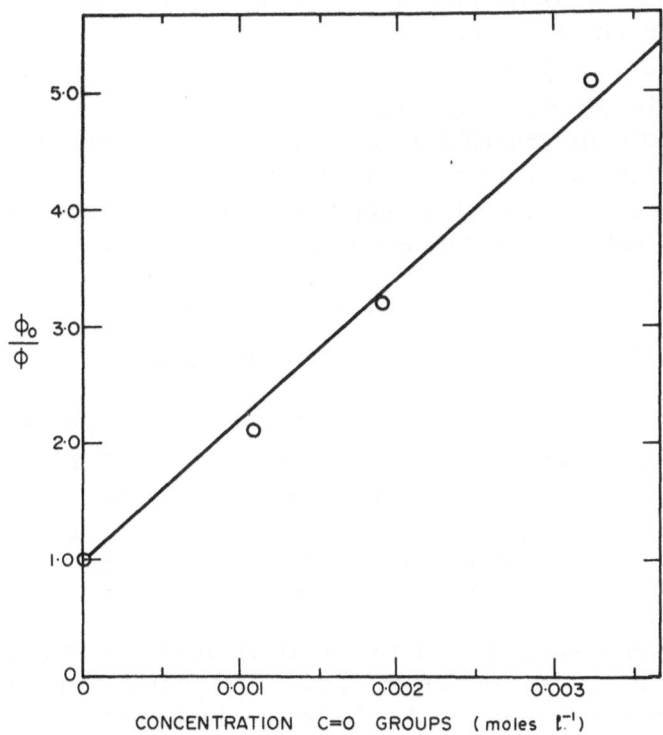

CONCENTRATION C=O GROUPS (moles ℓ^{-1})

mobility of the naphthalene in the solid matrix of the polymer. In fact if one assumes a constancy of the p factor, one can use these data to estimate k_{diff} and hence the diffusion coefficient D for naphthalene in polyethylene by substitution in the Smoluchowski equation:

$$k_{diff} = \frac{4\pi \rho DN}{1000}$$

where ρ is the radius of the diffusing molecule, D is the diffusion coefficient and N is Avogadro's number.

From this calculation a value of 1×10^8 ℓ $mole^{-1}$ sec^{-1} is obtained for the diffusion constant of naphthalene at 80° C. which compares very well with the value of 4×10^7 ℓ $mole^{-1}$ sec^{-1} determined for octadecane in low density polyethylene at the same temperature by radioactive tracer methods.

Lifetimes of Excited States

If we make the reasonable assumption that the efficiency factor p is similar for the two energy exchange processes considered, i.e.

$$^3(ketone) \longrightarrow {}^3(COD)$$
and $^1(naphthalene) \longrightarrow {}^1(ketone)$

then we may estimate the lifetimes of the triplet excited states of the ketone carbonyl in various environments from the experimental values of $k_q \tau$. The results of such calculations are summarized in Table II along with data obtained on low molecular weight ketones.

TABLE II

Triplet Lifetimes for Ketone Carbonyls

Ketone	Phase	Temp(°C)	(sec) x 10^9	Ref.
Ethylene–CO Copolymer	Solution	80	1.4	this work
	Solid film	80	10	"
6–undecanone	Solution	80	3.4	(7)
2–pentanone	Solution	25	5	(8)
2–hexanone	Solution	25	1	(8)

These results indicate that the lifetime of the carbonyl triplet in the polymeric ketone is very short, and of the same order of magnitude as in small molecules. This conclusion is supported by the earlier finding of Hartley and Guillet,[4] that atmospheric oxygen does not appear to quench any of the reaction and these workers concluded that, in order to observe quenching by oxygen at 90° C, the lifetime must be longer than 10^{-7} sec.

Conclusions

From these experiments it appears that the processes of electronic energy transfer in polymeric ketone systems both in solution and in the amorphous bulk phase above T_g are rather similar to those observed with smaller molecules. The lifetimes of the excited states appear to be of the same order of magnitude and the collisional efficiency for energy exchange nearly identical. Bimolecular processes involving reaction from excited states will be altered in polymer systems primarily by the difference expected in the diffusion rates of interacting species, particularly in the bulk phase.

The utility of energy transfer processes for the stabilization of polymers will depend to a large extent on the lifetime of the excited states involved. In the present instance involving aliphatic carbonyl groups, the excited state lifetimes, τ, are so short that excessive concentrations of the stabilizer molecules would be required to obtain an economically desirable effect.   However other polymers, particularly those containing phenyl groups might be expected to have longer lifetimes and hence would be easier to stabilize by this mechanism.   In view of this, the study of energy transfer processes in other polymers may well lead to important advances in the development of stabilizer systems.

The authors wish to acknowledge the financial support of the National Research Council of Canada and a fellowship (M.H.) from the Xerox Corporation.

BIBLIOGRAPHY

1.   M. Heskins and J.E. Guillet, Macromolecules, 1, 97 (1968).

2.   M. Heskins and J.E. Guillet, Polymer Preprints 9, 370 (1968).

3.   J.C.W. Chien and W.P. Connors, J. Amer. Chem. Soc., 90, 1001 (1968).

4.   G.H. Hartley and J.E. Guillet, Macromolecules, 1, 165 (1968).

5.   G.H. Hartley and J.E. Guillet, Macromolecules, 1, 413 (1968).

6.   R.S.H. Liu, J. Amer. Chem. Soc., 89, 112 (1967).

7.   M. Heskins and J.E. Guillet, Macromolecules (in press).

8.   P.J. Wagner and G.S. Hammond, J. Amer. Chem. Soc., 88, 1245 (1966).

FLUORESCENCE POLARIZATION KINETIC MEASUREMENTS OF ANTIGEN-ANTIBODY REACTIONS

S. A. Levison, F. Kierszenbaum and W. B. Dandliker[*]
This work is supported by The John A. Hartford Foundation, The National Science Foundation, The National Institutes of Health, (AM 7508, AM 5458 and Special Fellowhip AM 42568 from The National Institute of Arthritis and Metabolic Diseases.

SYMBOLS

Subscripts

e,	equilibrium value of parameter
f, b,	free and bound forms, respectively of fluorescent-labeled material
0,	at time approaching zero
(AB),	molar concentration of antibody
(AG),	molar concentration of antigen
F,	molar concentration of fluorescent-labeled material
$F_{b\,max}$,	concentration of combining sites in unlabeled component as determined by equation (3)
k,	defined by $-\dfrac{d\,(AG)}{dt} = k(AB)^{N1}\,(AG)^{N2}$ equation (7)
k_1,	bimolecular rate constant defined by equation (1)
k_{-1},	unimolecular rate constant defined by equation (1)
K,	$\dfrac{k_1}{k_{-1}}$ equals equilibrium association constant defined by equation (1)
k',	empirical rate constant defined by equation (10)
k'',	empirical rate constant defined by equation (11)
k_p,	unimolecular rate constant defined by equation (12)
N_1,	order of reaction with respect to (AB)
N_2,	order of reaction with respect to (AG)
p,	polarization of fluorescence
Q,	ratio of fluorescence intensity to molar concentration of fluorescent-labeled material
$(\dfrac{dp}{dt})$,	rate of change of polarization
K_o,	average association constant defined by equation (3)

* Division of Biochemistry
  Scripps Clinic & Research Foundation
  La Jolla, California 92037

INTRODUCTION
The polarization of fluorescent light from solutions has provided important information concerning the size, shape and conformation of macromolecules (1, 2), molecular anisostropy (3),
electronic energy transfer (4) and interactions which include dye
binding (5) to proteins. During the past few years the measurement
of both fluorescence polarization and intensity has been successfully utilized to determine both equilibrium (6, 7, 8, 9) and kinetic
parameters (10, 11, 12, 13, 14, 15) for antigen-antibody systems.
The basis of this approach involves the tagging of one of the reactants, e. g. the antigen with a small fluorescent molecule which is
then used as the detecting and measuring agent for its partner.
Changes in either the polarization of fluorescence intensity or the
fluorescence intensity itself can then be monitored directly and
hence afford a means by which the extent of reaction can be followed.
It is important to note that changes in the fluorescence polarization
parameter occur even in the absence of fluorescence quenching or
enhancement as long as there is a change in rotary brownian motion,
which results from the combination of the smaller fluorescent-la-
beled molecule with its larger unlabeled partner. Hence, fluorescence polarization measurements afford a powerful general approach
by which the kinetics and thermodynamics of important macromolecular reactions can be studied. This particular report, while including some thermodynamic data, centers mainly on the rates of reaction
between antigen and antibody molecules in the primary stages of
combination and on the effects of the ionic medium on these rates.

MATERIALS AND METHODS
Rabbit γG antibody and the corresponding univalent Fab fragments were prepared as described elsewhere (16, 17). Both the
preparation and purification of fluorescent labeled antigens have
also been described previously (16, 18). Narrow fractions of each
were obtained from the center portions of the peaks. The dye to
protein molar ratios for dansyl-BSA was about 2. 7 whereas for
fluorescein-ovalbumin it was about 0. 7. The test antigen for antifluorescein activity was γG labeled with fluorescein having a dye to
protein ratio of 0. 45. Precipitin determinations were used to measure divalent antibody concentrations (10). Thermodynamic measurements were carried out with a previously described fluorescence
polarometer (8) whereas, the kinetic studies were implemented
with a direct readout fluorescence polarometer (19) having a response time of about 5 sec. The delivery and mixing procedures
were accomplished in less than 10 sec. During the course of reaction, the cell contents were maintained at $1.5 \pm 0.5°$.

EQUATIONS
Both thermodynamic and kinetic equations have been developed for systems in which it is assumed that a fluorescent-tagged
antigen molecule, $\mathcal{F}$, reacts reversibly with an antibody, $\mathcal{R}$;

$$\overset{\alpha}{\mathcal{F}} + R \;\; \underset{k_{-1}}{\overset{k_f}{\rightleftharpoons}} \;\; \mathcal{F}R \tag{1}$$

Formulation of these equations were based upon the following expression which relates the concentration of free fluorescent antigen, F_f, to the bound form F_b (8).

$$\frac{F_b}{F_f} = \frac{Q_f}{Q_b}\left(\frac{p - p_f}{p_b - p}\right) \tag{2}$$

Equilibrium Method

The following expression for the mass law has been applied to antigen-antibody systems in which the binding sites are heterogeneous and their distribution is 'compatible' with a Sips distribution of binding sites (8, 9, 16):

$$\log F_f = \frac{1}{a}\left[\log\left(\frac{F_b}{F_{b\,max} - F_b}\right)\right] - \log K_0 \tag{3}$$

(if the sites are homogeneous, a, the heterogeneity constant equals one).

Kinetic Techniques

Kinetic methods employing initial rate equations as well as the integrated rate expression have been developed directly in terms of fluorescence polarization and intensity parameters: The initial rate method has been most useful in ascertaining the form of the empirical rate law by determining the order of reaction with respect to each reactant. The initial rate of change of polarization as previously formulated is (10):

$$\left(\frac{dp}{dt}\right)_0 = \frac{Q_b}{Q_f}(p_b - p_f)\,k\,(AB)_0^{N_1}(AG)_0^{N_2 - 1} \tag{4}$$

where k is the usual empirical rate constant. For constant $(AB)_0$ but varying $(AG)_0$,

$$\log\left[\left(\frac{dp}{dt}\right)_0\right] = (N_2 - 1)\log\left[(AG)_0\right] + \text{constant}_a \tag{5}$$

where constant $_a$ = $\log\left[\frac{Q_b}{Q_f}(p_b - p_f)\,k\,(AB)_0^{N_1}\right]$

For constant $(AG)_0$ but varying $(AB)_0$,

$$\log\left[\left(\frac{dp}{dt}\right)_0\right] = N_1 \log\left[(AB)_0\right] + \text{constant}_b \tag{6}$$

where constant $_b$ = $\log \left[\frac{Q_b}{Q_f} (P_b - P_f) k (AG)_b^{N_2-1} \right]$

$(\frac{dp}{dt})_o$ refers to the rate of change of polarization as $t \longrightarrow$ o.
N_1 and N_2 are the orders of reaction with respect to (AB) and (AG)
respectively and have been defined by a rate law of the following
form (20):

$$-\left(\frac{d (AG)}{dt} \right)_{t \to o} = k (AB)_o^{N_1} (AG)_o^{N_2} \quad (7)$$

If the reaction rate is simple second order, then the integrated
rate method provides a means by which the values of k_1 and k_{-1}
and hence K, the equilibrium association constant, (see equation (1))
can be determined. The method also serves as a useful check on
initial rate results. The integrated rate expression applicable to
equation (1) can be formulated in terms of fluorescence polarization
parameters as a pseudo first-order equation if the antibody concentration is in such excess that it is constant during the course of
reaction (10, 12):

$$\log \left[\frac{\frac{(P_e - P_f)}{P_e (Q_f/Q_b - 1) + (P_b - P_f)}}{\frac{(P_e - P_f)}{P_e (Q_f/Q_b - 1) + (P_b - P_f)} - \frac{(P - P_f)}{P (Q_f/Q_b - 1) + (P_b - P_f)}} \right] = \left(\frac{k_1' + k_{-1}}{2.3} \right) t \quad (8)$$

where $k'_1 = k_1 (AB)^{N_1}$. If $Q_f/Q_b = 1$, equation (8) simplifies to

$$\log (P_e - P) = \log (P_e - P_f) - \left(\frac{k_1 (AB)_o^{N_1} + k_{-1}}{2.3} \right) t \quad (9)$$

Hence, a plot of log $(p_e - p)$ vs time should be linear with a slope
equal to $-(k_1 (AB)_o^{N_1} + k_{-1})/ 2.3$ (denoted by S). Furthermore,
a plot of S vs $(AB)_o$ will be linear provided that the order with respect to (AB) is one. The rate constants k_1 and k_{-1} are obtained
from the slope and intercept respectively of the latter plot.

RESULTS

Thermodynamics
Equilibrium studies of a number of immunochemical reactions
with ovalbumin, BSA and γ-globulin as antigens and with the peni-
cilloyl, dansyl, and fluorescein groups as haptens have been
studied. Equation (3) has been used to determine such parameters
as binding site concentration, heterogeneity constant and association constant for the above systems. A summary of such parameters
is given in Table I.

Table I: Some Thermodynamic Parameters for Various Antigen, Antibody Systems[a]

Antibody	Anti FO[b]	Fab	Antiovalbumin	Anti FO	Anti BSA[c]	Human Antipenicillin	Rabbit Antipenicillin
Antigen	FO[b]	FO	FO	F[d]	dansylBSA[c]	PD AB-F[e]	PDABF
$10^8 F_{b\,max}^f$	5.0	2.5	1.5	13.5	2.0	1.3	4.3
$10^{-8} K_o^f$	2.1	2.6	1.8	0.054	2.0	0.3	0.087
a	0.84	1.1	0.65	0.51	0.61	0.78	0.71
Reference	8	8	8	8	16	9	9

a   All titrations were carried in 0.15 M NaCl, 0.01 M Na_2HPO_4, 0.005 M Na_2HPO_4 at $23°$.

b   Anti FO and FO denote antifluorescein labeled ovalbumin and fluorescein-labeled ovalbumin respectively.

c   AntiBSA and dansyl BSA denote antibovine serum albumin and dansyl-labeled bovine serum albumin.

d   F refers to fluorescein.

e   PD AB-F refers to penicillin 1, 4 diaminobutane fluorescein.

f   All concentrations are expressed in molarity. Thus $F_{b\,max}$ is in molarity of sites capable of binding the antigen or hapten in question and K_o is in reciprocal molarity.

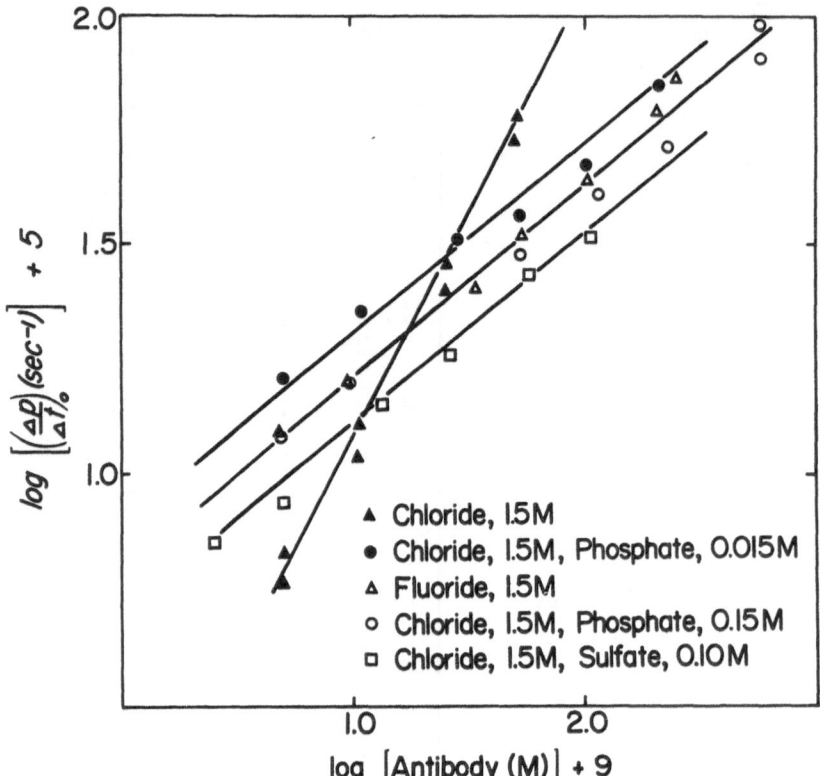

Figure 1. Effect of non-chaotropic ions on the order of reaction with respect to antibody concentration for the fluorescein-labeled ovalbumin divalent antiovalbumin system (see equation (6)). All of the studies were performed in pH 7.0 solution at $1.5 \pm 0.5°$. The antigen was a narrow fraction of fluorescein-labeled ovalbumin, purified on Sephadex G-100. The antibody was an IgG preparation isolated by DEAE cellulose chromatography. Similar results were obtained with immunospecifically purified antibody (12). ▲-1.5 M KCl, 0.01 M Tris; ☐ -1.5 M KCl, 0.1 M K_2SO_4, 0.01 M Tris; △-1.5 M KF, 0.01 M Tris; O-1.5 M KCl, 0.01 M K_2HPO_4, 0.005 M KH_2PO_4; O - 1.5 M KCl, 0.1 M K_2HPO_4, 0.05 M KH_2PO_4.

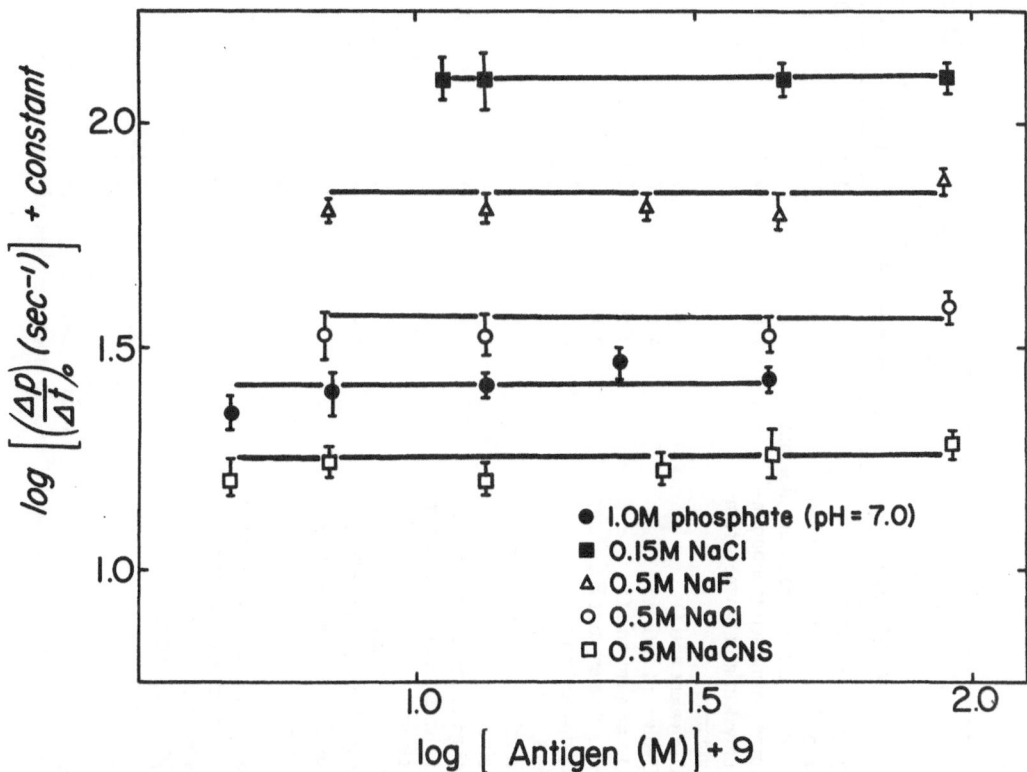

Figure 2. Determination of the order of reaction with respect to antigen for the dansyl-labeled bovine serum albumin (BSA), divalent anti BSA system in various ionic media at $1.5 \pm 0.5^{\circ}$ (see equation (5)). All solutions were buffered at pH 7.0. ● - 1.0 M phosphate; ■ - 0.15 M NaCl, 0.01 M Tris; △ - 0.5 M NaF, 0.01 M Tris; O - 0.5 M NaCl, 0.01 M Tris; □ - 0.5 M NaCNS, 0.01 M Tris.

Medium (M)	AG–AB System	AB Order, N_1	AG Order, N_2	Factor of Concn Change (AB)	(AG)
NaCNS (1.5)	FO[b]–divalent anti O	1.0	1.0	10	10
NaCNS (0.5)	Dansyl bovine serum albumin–divalent anti bovine serum albumin	0.95	1.0	20	20
NaCNS (0.5)	Dansyl bovine serum albumin–univalent anti bovine serum albumin	1.0	1.0	14	20
NaCNS (0.15)	FO–divalent anti O	1.0	1.0	8	16
NaClO$_4$ (1.5)	FO–divalent anti O	1.0	0.95	16	10
NaClO$_4$ (0.15)	FO–univalent anti O	1.0	1.0	5	5
NaClO$_4$ (0.15)	FγG,[c] divalent anti F	1.0	1.0	10	11
NaCl (1.5)	FO–divalent anti O	0.97	1.1	10	10
NaCl (0.5)	Dansyl bovine serum albumin–divalent anti bovine serum albumin	1.0	1.0	10	20
NaCl (0.5)	Dansyl bovine serum albumin–univalent anti bovine serum albumin	0.93	1.0	20	14
NaCl (0.15)	FO–divalent anti O	1.0	1.0	20	10
NaCl (0.15)	Dansyl bovine serum albumin–divalent anti bovine serum albumin	1.0	1.0	20	10
NaCl (0.15)	FγG–divalent anti F	1.0	1.0	10	10
NaCl (0.15)	Dansyl bovine serum albumin–univalent anti bovine serum albumin	1.0	1.0	20	10
NaCl (0.15)	FγG–univalent anti F	1.0	1.0	10	10
KCl (1.5)	FO–divalent anti O	1.0	1.0	10	50
KCl (0.15)	FO–divalent anti O	0.93	1.0	10	10

[a] All experiments were carried out in 0.01 M Tris buffer, pH 7.0, at 1.5 ± 0.5°. [b] FO and anti O denote fluorescein-labeled ovalbumin and antiovalbumin, respectively. [c] FγG and anti F refer to fluorescein-labeled γ-globulin and antifluorescein.

Medium (M)	AG–AB System	AB Order, N_1	AG Order, N_2	Factor of Concn Change	
				(AB)	(AG)
KCl (1.5)	FO[b]–divalent anti O	0.44	1.0	50	50
K₂HPO₄ (0.1)	FγG[c]–divalent anti F	0.53	1.0	10	10
KH₂PO₄ (0.05)	FO–univalent anti O	1.0	1.0	10	5
	FγG–univalent anti F	1.0	1.0	10	7
K₂HPO₄ (0.67)	FO–divalent anti O	0.47	1.0	80	20
KH₂PO₄ (0.33)	Dansyl bovine serum albumin–divalent anti bovine serum albumin	0.45	1.0	40	10
	Dansyl bovine serum albumin–univalent anti bovine serum albumin	1.0		40	10
KCl (1.5) K₂HPO₄ (0.01) KH₂PO₄ (0.005)	FO–divalent anti O	0.46	1.0	40	10
NaCl (1.5) Na₂HPO₄ (0.01) NaH₂PO₄ (0.005)	FO–divalent anti O	0.5	1.0	10	10
NaCl (0.15)	FO–divalent anti O	0.61	1.0	200	100
Na₂HPO₄ (0.01)	Dansyl bovine serum albumin–divalent anti bovine serum albumin	0.8	1.0	10	13
Na₂H2–PO₄ (0.005)	FO–univalent anti O	0.92	1.0	30	100
KCl[d] (1.5) K₂SO₄ (0.1)	FO–divalent anti O	0.45	1.0	40	20
KF[d] (1.5)	FO–divalent anti O	0.48	0.93	40	10
KF (1.5)	FO–univalent anti O	1.0	1.0	10	5
NaF[d] (0.5)	Dansyl bovine serum albumin–divalent anti bovine serum albumin	0.6	1.0	40	13
NaF[d] (0.5)	Dansyl bovine serum albumin–univalent anti bovine serum albumin	1.0	1.0	20	13

[a] All experiments were carried out at 1.5 ± 0.5°. [b] FO and anti O denote fluorescein-labeled ovalbumin and antiovalbumin, respectively. [c] FγG and anti F refer to fluorescein-labeled γ-globulin and antifluorescein, respectively. [d] Buffered with 0.01 M Tris at pH 7.0.

Kinetics

Initial rate measurements employing equations (5) and (6) have been utilized to determine the empirical rate law in various ionic media for three different antigen-antibody systems: fluorescein labeled ovalbumin, antiovalbumin; dansyl-labeled BSA, anti BSA; and fluorescein-labeled γG, antifluorescein. In all three cases, it has been shown that the initial rate of reaction is markedly influenced by the nature of the ionic medium and that an interesting relationship between the Hofmeister series and the initial rate law exists. Specifically, for antigen, divalent antibody systems, chaotropic ions, (e. g., chloride, perchlorate and thiocyanate) which promote macromolecular unfolding and dissociation correlate with a simple second-order rate law:

$$-\left(\frac{d\,(AG)_0}{dt}\right)_0 = k'\,(AB)\,(AG) \tag{10}$$

whereas non-chaotropic ions, which promote macromolecular folding and association correlate with a more complicated initial rate law involving a fractional order with respect to antibody concentration:

$$-\left(\frac{d\,(AG)_0}{dt}\right)_0 = k''\,(AB)^{N_1}\,(AG) \tag{11}$$

where N_1 is usually close to $1/2$. To determine N_1, the order of reaction with respect to divalent antiovalbumin, plots of $\log\left[(dp/dt)_0\right]$ vs $\log\left[(AB)_0\right]$ at constant $(AG)_0$ were made as shown in Figure 1. Examples of the determination of the order with respect to divalent anti BSA are shown in Figure 2, where $\log\left[(dp/dt)_0\right]$ is plotted against $\log\left[(AG)_0\right]$ at constant $(AB)_0$. Further initial rate studies have also been made on the binding of antigen to univalent antibody (Fab) for the three previously mentioned immunochemical systems. The results differ from those obtained for divalent antibody systems in that the reaction rate for the antigen, univalent antibody systems in all the ionic media obey equation (10), that is, simple second-order kinetics. A summary of the determined orders, N_1 and N_2 is presented in Tables II and III.

To confirm the initial rate results as well as to determine specific salt effects on the back reaction, the left-hand side of equation (9) was plotted vs time for different initial concentrations of antibody. The plots were linear at least up to the half time of polarization change. The deviation from pseudo first-order kinetics may be indicative of site non-uniformity, site depletion or possibly of a change in the reaction mechanism. The initial slope $S, \left(k_1\,(AB)_0 + k_{-1}\right)/2.3$, was then plotted as a function of $(AB)_0$ as shown in Figures 3 and 4 for the fluorescein-labeled ovalbumin, divalent antiovalbumin and the dansyl-labeled bovine serum albumin, univalent anti BSA systems, respectively. The magnitudes of k_1 and k_{-1} as well as the k values obtained by initial rate measurements are listed in Tables IV and V.

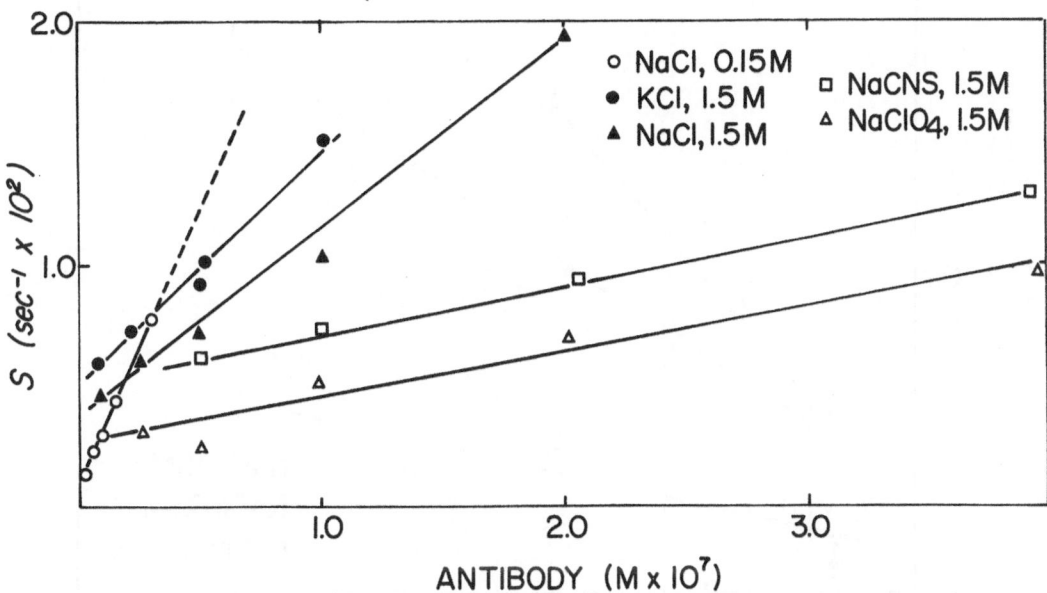

Figure 3. Plots of pseudo first-order rate constants vs. divalent antiovalbumin concentration in various ionic media. S denotes pseudo first-order parameter, $\left(k_1(AB)\ k_{-1}\right)$ / 2.3 (see equation (9)). The slope and intercept of these plots yield the second-order dissociation constant, k_1, and the first-order dissociation constant, k_{-1}, respectively, for the ovalbumin-antiovalbumin reaction. All experiments were performed in pH 7.0, 0.01 M Tris at $1.5 \pm 0.5°$. Antigen-antibody preparations were the same as those of Figure 1. O - 0.15 M NaCl; ● - 1.5 M KCl; ▲ - 1.5 M NaCl; ☐ - 1.5 M NaCNS; △ -1.5 M NaClO$_4$.

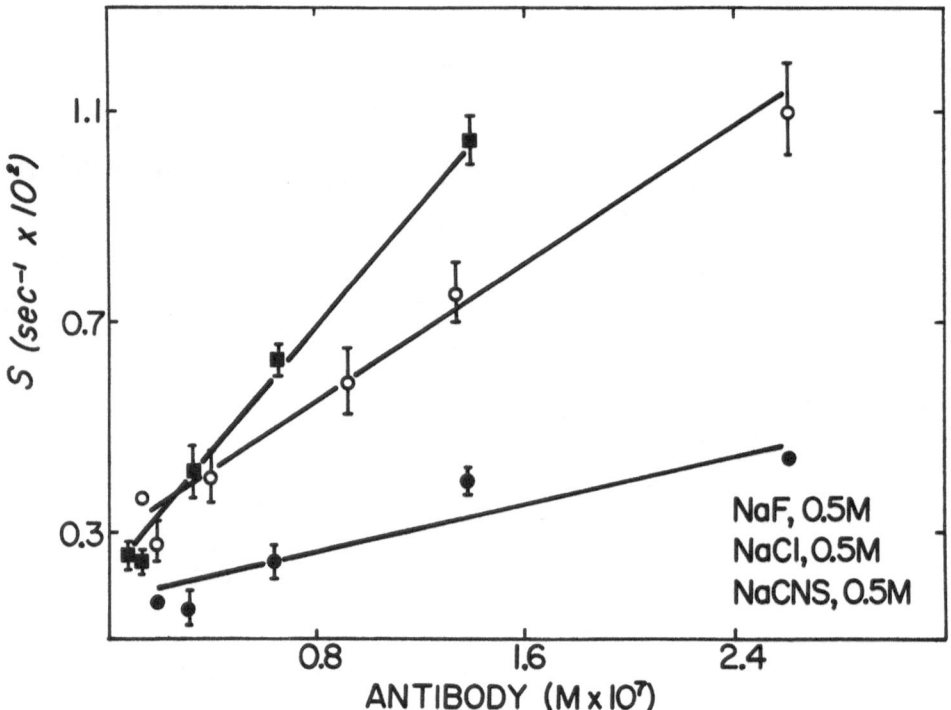

Figure 4. Specific ion effects on the reaction kinetics of the dansyl-labeled bovine serum albumin (BSA), univalent antiBSA system at $1.5 \pm 0.5^{\circ}$; $\Gamma/2 = 0.50$. S denotes pseudo first-order parameter, $\left(k_1 (AB) \; k_{-1}\right) / 2.3$ (see equation (9)). The slope and intercept of these plots yield the second-order association constant, k_1, and the first-order dissociation constant, k_{-1}. All experiments were performed in 0.01 M pH 7.0 Tris buffer. ■ - NaF;   O - NaCl;   ● - NaCNS.

TABLE IV: Summary of Rate and Equilibrium Constants for the Fluorescein-Labeled Ovalbumin–Antiovalbumin and the Fluorescein-Labeled γ-Globulin–Antifluorescein Systems in Various Ionic Media.[a]

Medium (M)	k_1 (M^{-1} sec^{-1}) × 10^{-4}[b]	k_{-1} (sec^{-1}) × 10^{2}[b]	K (M^{-1}) × 10^{-6}[c]	Initial Rate Constant k[d] (M^{-1} sec^{-1}) × 10^{-4}
FO–divalent anti O				
NaCNS[f] (3.0)				10^{-2}–10^{-1}
NaCNS (1.5)	4.0	1.2–1.0	3.3–4.0	3.8 ± 0.4
NaClO$_4$ (1.5)	3.7	0.5–1.0	3.7–7.4	2.8 ± 0.2
NaCl (1.5)	18	0.8	23	15 ± 3
KCl (1.5)	23	1.1	21	24 ± 2
NaCNS (0.15)	29	0.5–1.0	30–60	20 ± 2
NaCl (0.15)	48	0.2	240	50 ± 6
KCl (0.15)				68 ± 8
FO, univalent anti O				
NaCl (0.15)	20	0.8	25	15 ± 3
Na$_2$HPO$_3$ (0.01) } NaH$_2$PO$_4$ (0.005) }				
KF (1.5)	12	0.5	24	
NaClO$_4$ (0.15)	11	0.5	22	10 ± 3
KCl (1.5) } K$_2$HPO$_4$ (0.1) } KH$_2$PO$_4$ (0.05) }	8	0.7	1.1	8 ± 1
FγG,[e] divalent F				
NaCl (0.15)				62 ± 4
FγG, univalent anti F				
KCl (1.5) } K$_2$HPO$_4$ (0.1) } KH$_2$PO$_4$ (0.05) }				3 ± 0.2
NaCl (0.15)				15 ± 1

[a] All experiments were carried out at 1.5 ± 0.5° in solutions buffered at pH 7.0. [b] Defined by eq 6. [c] $K = k_1/k_{-1}$. [d] Defined by eq 4. Typical $p_b - p_f$ and Qf/Qb values used to evaluate k for the anti O systems were 0.05 and 1.0, respectively. Values of $p_b - p_f$ and Qf/Qb values used for the divalent anti F systems were 0.13 and 1.8, respectively, while $p_b - p_f$ and Qf/Qb values used for the univalent anti F systems were 0.14 and 2.9, respectively. [e] FO and anti O denote fluorescein-labeled ovalbumin and antiovalbumin, respectively. [f] All solutions except those which are explicitly listed as containing phosphate buffer were buffered with 0.01 M Tris. [g] FγG and anti F refer to fluorescein-labeled γ-globulin and antifluorescein, respectively.

TABLE V: Summary of Rate and Equilibrium Constants for the Dansyl-Labeled Bovine Serum Albumin–Anti Bovine Serum Albumin System in Various Ionic Media.[a]

Medium (M)	k_1 (M⁻¹ sec⁻¹) × 10^{-4} [b]	k_{-1} (sec⁻¹) × 10^{2} [b]	K (M⁻¹) × 10^{-6} [c]	Initial Rate, k (M⁻¹ sec⁻¹) × 10^{-4} [d]
Dansyl bovine serum albumin–divalent anti bovine serum albumin				
NaCl[e] (0.15)	34	0.2	170	32 ± 4
NaCl (0.5)	11	0.7	16	11 ± 2
NaCNS (0.5)	7.9	0.6	13	6.8 ± 2
Dansyl bovine serum albumin–univalent anti bovine serum albumin				
NaCl (0.15)	11	1.0	11	10 ± 1.2
K₂HPO₄ (0.67) KH₂PO₄ (0.35)	10	0.5	20	11 ± 1.5
NaCl (0.5)	7.3	0.7	10	7.5 ± 1.0
NaCNS (0.5)	2.0	0.6	3	1.9 ± 0.2
NaF (0.5)	13.0	0.6	21	10 ± 1.0

[a] All studies were carried out at 1.5 ± 0.5° in solutions buffered at pH 7.0. [b] Defined by eq 5. [c] $K = k_1/k_{-1}$. [d] Defined by eq 4. Typical values of $p_b - p_f$ and Q_f/Q_b used to evaluate k were 0.06 and 1.0, respectively. [e] All solutions except those which are explicitly listed as containing phosphate were buffered with 0.01 M Tris.

DISCUSSION

This investigation has reported on both thermodynamic and kinetic studies of the primary antigen-antibody combination. By coupling the techniques of fluorescence labeling to those of classical chemical kinetic techniques, a general and direct means for both thermodynamic and kinetic studies of macromolecules has been described. This particular report has especially focussed on the effects of the ionic medium upon the kinetics of the primary combination between antigen and antibody. Initial rate studies of three different immunochemical systems: ovalbumin, antiovalbumin; BSA, anti BSA and fluorescein γG, antifluorescein have shown the empirical rate law to be simple second-order in chaotropic media, i.e. first-order with respect to $(AB)_o$ and first-order with respect to $(AG)_o$, whereas in non-chaotropic media, the reaction rate involves a fractional order (usually close to 1/2) with respect to (AB). It has also been shown that the primary combination between antigen and its corresponding univalent antibody partner, Fab, obeys simple second-order kinetics in all ionic media investigated. Hence, intact antibody and Fab antibody must react with antigen by different processes in non-chaotropic media. Finally, there is a correlation with the magnitude of all the second-order rate constants determined in this study with the Hofmeister series, that is, the value of the second-order constant in different ionic environments varies according to the following sequences of ions:

$$ SCN^- \langle ClO_4^- \langle Cl^- \langle F^- \langle SO_4^- \langle phosphate $$

These specific anion effects on the binding kinetics of antigen-antibody systems follow similar trends established in studies involving macromolecular stability (21), enzymatic steady state reactions (22, 23) and the dissociation of both hapten, antihapten (24) and antigen-antibody complexes (25). Generally speaking, the chaotropic anions such as SCN^-, ClO_4^-, and Cl^- which are at one end of the Hofmeister series tend to unfold and dissociate macromolecules whereas anions which are at the other end, (phosphate, sulfate and fluoride) tend to promote folding and association. It is particularly interesting that these effects extend even to the realm of small molecule interactions since they have been noted in the study of salt effects on SN_2 reactions between 2,4-dinitrochlorobenzene with aniline, hydroxide or thiophenoxide (26). The most reasonable explanation of these salt effects on antigen-antibody kinetics seems to be in terms of the solvation and conformational changes which occur when antigen and antibody molecules combine and we will now develop this interpretation.

Plausible Reaction Mechanisms

The simplest mechanism consistent with the observed facts that antigen, divalent antibody reactions in chaotropic media, and antigen, Fab reactions in all ionic media obey second-order kinetics in the initial stages of reaction is one which involves a bimolecular combination between antigen and antibody:

$$AB + AG \xrightarrow{k_1} AB\text{-}AG \qquad (12)$$

The fact that the magnitudes of the second-order rate constants are particularly sensitive to the nature of the specific anion present implies that the bimolecular process involved in equation (12) is not diffusion controlled but involves important structural changes. A recent determination of an activation energy of 12kcal/mol for the ovalbumin, divalent antiovalbumin combination by fluorescence polarization kinetic techniques (11) as well as interpretations of electromicrographs support this viewpoint (27, 28). It has been suggested that the divalent antibody molecule during its combination with its hapten or antigen partner clicks open around a central hinge and acquires an open rod-like shape, as compared to its more globular shape prior to reaction. It is our view that such conformational changes probably involve solvent reorganization and solvent loss, and or unfolding during formation of the activated complex. The idea of solvent loss is suggested by the fact that estimations of the entropy of activation for the ovalbumin, divalent antiovalbumin reaction in 0.15 M Na Cl at neutral pH yields an unusual, high positive value (11) when compared to other bimolecular associations (29) as well as to the hapten, antihapten reaction (30). Furthermore, if the activated complex is less solvated than the isolated reactants, then anions which compete more effectively for solvent molecules will tend to promote the reaction rate. This is precisely what is observed experimentally. Anions with high charge density tend to enhance the reaction rate whereas anions with low charge density tend to depress the reaction. *

To account for the fractional order observed for antigen, divalent antibody kinetics in non-chaotropic media, we have proposed a mechanism in which there is rapid reversible formation of a loosely held encounter pair of antigen and antibody macromolecules. A slower, unimolecular process in which the final antigen-antibody complex is formed then follows (12):

$$AB + AG \xrightleftharpoons{\text{fast}} AB \text{ --- } AG \qquad (11)$$

$$AB \text{ --- } AG \xrightarrow[\text{slow}]{k_p} AB\text{—}AG \qquad (12)$$

It is also assumed that the reacting sites exhibit a non-uniformity which is manifested in the encounter pair formation and that the encounter pair formation equilibrium involves a Sips distribution of binding energies (32). The initial rapid interaction between antigen and antibody may be thought of as relatively weak and occurring at a few contact points (possible near or at one of the antibody specific sites) in which hydrophobic and or electrostatic bonding occurs. Many solvent molecules would still be interdispersed between the antigen and antibody molecules which make-up the loose encounter pair. The rate determining step yielding the final product would involve a gradual realignment of antigen and

*An equivalent analysis of these results has been made in terms of hydrophobic bonding and its relation to chaotropic ions (31).

antibody bonds in which the second valence of the antibody molecule may play an important role and in which solvent molecules may be released either as the result of neutralization of charge and or of hydrophobic bonding.

The proposed macromolecular encounter pair in many ways resembles the ion-pair whose chemical behavior is dependent on whether the reactants are in intimate contact or solvent separated (33, 34, 35, 36,). In fact, the above described encounter pair mechanism is quite similar to the general mechanism of complexation of ligand to metal ions which involves formation of an ion pair followed by a rate determining dissociation of one or more water molecules (34, 35). However, it should be pointed out that the initial encounter pair formation unlike ion-pair reactions is not necessarily diffusion controlled. Furthermore, the macromolecular encounter pair stability depends not only on electrostatic forces as in ion pairing, but also on apolar interactions.

The second-order rate constants obtained for divalent antibody systems in chaotropic media suggest that only one of the antibody valences may be kinetically important and that the rearrangement of the encounter pair (equation 12) occurs very rapidly compared to the equilibrium involving encounter pair formation (equation 11). This idea is supported by the fact that Fab (univalent) fragments in all media obey second-order kinetics (13) and they behave hydrodynamically as simple globular proteins as opposed to the much more flexible divalent form (37). One would also expect that the divalent form with its high degree of flexibility should be much more sensitive to intramolecular folding effects (caused by the non-chaotropic high charge density anions) than the more globular univalent form. It may be that intramolecular folding is in part responsible for the non-uniformity of sites in formation of the encounter pair. In these terms, the kinetic effects of intramolecular folding of the antibody molecule can be minimized either by splitting whole antibody into Fab fragments or by changing the environemnt from a non-chaotropic to a chaotropic medium where hydrophobic bonds are broken. Finally, the encounter-pair mechanism can also be thought of in terms of a model previously presented to explain fractional order rate laws which were observed in heterogeneous catalysis (38). The key idea is that for systems which have a non-uniformity of binding sites, having a complete available range of activation energies, there is no single rate-determining step but rather different rate determining steps on different types of sites. On some sites, encounter pair formation (equation (11)) would be rate limiting whereas on others, rearrangement of the encounter pair (equation (12)) would be rate limiting. On a third set, both processes would be equally slow. It is assumed, no matter what set of sites are considered, that the sum of the two activation energies for any site is always equal to a constant overall activation energy. This last statement implies that on one kind of site, slow encounter pair formation (high activation energy), would lead to rapid rearrangement (low activation energy), and rapid encounter pair formation (low activation energy) would necessitate a slow rearrangement (high activation energy).

Acknowledgements

     The technical assistance of Maria Vila is gratefully acknowledged.

Tables II - V and Figures 1 - 4 are reproduced from Biochemistry 9, 322-331 (1970) by S. A. Levison, F. Kierszenbaum and W. B. Dandliker, with the permission of the American Chemical Society.

BIBLIOGRAPHY

1. Weber, G. Biochem. J., 51, 145 (1952).
2. Steiner, R. F. and Edelhoch, H. Chem. Rev., 62, 457 (1962).
3. Albrecht, A. J. Mol. Spectroscopy, 6, 84 (1961).
4. Weber, G. and Teale, F. W. J. The Proteins, 3, 445, edited by Neurath, Academic Press, New York (1965).
5. Laurence, D. J. R. Biochem. J., 51, 168 (1952).
6. Dandliker, W. and Feigen, G. Biochem. Biophys. Res. Comm., 5, 299 (1961).
7. Haber, E. and Bennett, J. C. Proc. Nat. Acad. Sci., 48, 1935 (1962).
8. Dandliker, W. B., Schapiro, H. C., Meduski, J. W., Alonso, R., Feigen, G. A. and Hamrick, J. R. Jr. Immunochem., 1, 165 (1964).
9. Dandliker, W. B., Halbert, S. P., Florin, M. C., Alonso, R. and Schapiro, H. C. J. Exp. Med., 122, 1029 (1965).
10. Dandliker, W. B. and Levison, S. A. Immunochem., 5, 171 (1967).
11. Levison, S. A., Jancsi, A. N. and Dandliker, W. B. Biochem. Biophys. Res. Comm. 33, 942 (1968).
12. Levison, S. A. and Dandliker, W. B. Immunochem., 6, 253 (1969).
13. Levison, S. A., Kierszembaum, F. and Dandliker, W. B. Fed. Proc., 28, 326 (1969).
14. Tengerdy, R. P. Immunochem., 3, 463 (1966).
15. Tengerdy, R. P. J. Immunol., 99, 126 (1967).
16. Kierszenbaum, F., Dandliker, J. and Dandliker, W. B. Immunochem., 6, 125 (1969).
17. Porter, R. R. Biochem. J., 73, 119 (1959).
18. Kierszenbaum, F., Levison, S. A. and Dandliker, W. B. Anal. Biochem., 28, 563 (1969).
19. White, J. U., Williamson, D. E., Levison, S. A. and Dandliker, W. B. (in preparation).

20. Frost, A. A. and Pearson, R. G. Kinetics and Mechanism, p. 186, John Wiley, Inc. , New York (1961).
21. von Hippel, P. H. and Wong, K. Y. Science, $\underline{145}$, 577 (1964).
22. Warren, H. C. and Cheatum, S. G. Biochem. , $\underline{5}$, 1702 (1966).
23. Warren, J. C. , Stowring, L. and Morales, M. F. J. Biol. Chem. , $\underline{241}$, 309 (1966).
24. Pressman, D. , Nissonoff, A. and Radzimski, G. J. Immunol. , $\underline{86}$, 35 (1961).
25. Dandliker, W. B. , Alonso, R. , de Saussure, V. A. , Kierszenbaum, F. , Levison, S. A. and Schapiro, H. C. Biochem. , $\underline{6}$, 1460 (1967).
26. Bunton, C. A. and Robinson, L. J. Am. Chem. Soc. , $\underline{90}$, 5965 (1968).
27. Feinstein, A. and Rowe, A. J. Nature, $\underline{205}$, 147 (1965).
28. Valentine, K. and Green, N. J. Molec. Biol. , $\underline{27}$, 615 (1967).
29. Laidler, K. J. Chemical Kinetics, p. 198. McGraw-Hill, Inc. , New York, N. Y. (1965).
30. Day, L. A. , Sturtevant, J. M. and Singer, S. J. Ann. N. Y. Acad. Sciences, $\underline{103}$, 611 (1963).
31. Dandliker, W. B. and de Saussure, V. A. In The Chemistry of Biosurfaces, Edited by M. Hair, Marcel Dekker, N. Y. (1970).
32. Sips, R. J. Chem. Phys. , $\underline{16}$, 490 (1948).
33. Winstein. S. , Appel, B. , Baker, R. and Diaz, A. The Chemical Society, London, Special Publication No. 19, pp. 109-130, (1965).
34. Eigen, M. Z. Elektrochem. , $\underline{64}$, 115 (1960).
35. Eigen, M. and Tamm, K. A. Elektrochem. , $\underline{66}$, 107 (1960).
36. Szwarc, M. Accounts of Chemical Research, $\underline{2}$, 87 (1968).
37. Noelken, M. E. , Nelson, C. A. , Buckley, C. E. III and Tanford, C. J. Biol. Chem. , $\underline{240}$, 218 (1965).
38. Halsey, G. D. J. Chem. Phys. , $\underline{17}$, 758 (1949).

THE DYE-SENSITIZED PHOTOOXIDATION OF BIOLOGICAL MACROMOLECULES*

John D. Spikes and Martha L. MacKnight

Department of Biology, University of Utah

Salt Lake City, Utah

INTRODUCTION

Almost all kinds of biological macromolecules,including polysaccharides, nucleic acids and proteins, are oxidized upon illumination in the presence of sensitizing dyes and molecular oxygen.
This phenomenon is termed "photodynamic action" by biologists (1).
Although the degradation by light of dyed fabrics composed of cotton, silk, wool, etc., was known even in early times, the first
formal description of the sensitized photooxidation of biological
macromolecules was made by Professor von Tappeiner and his students
in Munich in 1903. They reported that proteins (enzymes) such as
diastase, invertase, papain and trypsin were inactivated on aerobic
illumination in the presence of eosin (2). Since that time, photodynamic studies have been made on a large number of different proteins, mainly crystalline enzymes, as well as on polysaccharides
and nucleic acids.

MECHANISMS OF SENSITIZED PHOTOOXIDATION REACTIONS

The sensitized photooxidation of biological macromolecules
appears to proceed by two major mechanisms, which may occur separately or simultaneously, depending upon the system studied. The
relative participation of the two pathways depends on the sensitizing dye used, the substrate (the biological molecule photooxi-

---

*The preparation of this paper was supported by the U. S. Atomic
Energy Commission under Contract No. AT(11-1)-875 and by National
Science Foundation Graduate Fellowship stipends to M. L. M.

67

dized), and the reaction conditions (reactant concentrations, pH, solvent composition, etc.). In general, photosensitized oxidations using molecular oxygen probably involve the triplet or some other metastable excited state of the dye (3). Dyes which do not produce long-lived states on illumination in solution, such as the triphenylmethane dyes, do not usually sensitize photooxidation reactions (4). Also, for example, 2,6-anthraquinone disulfonate, an efficient sensitizer for the photooxidation of alcohols, gives a good population of long-lived states in flash photolysis studies, while the 1,5-disulfonate, a poor photosensitizer, shows no long-lived states (5).

In photodynamic systems where the solvent is non-reactive and where dye concentrations are sufficiently low such that dye-dye reactions are minimized, the triplet dye will then react with oxygen and/or substrate, with the relative importance of these alternate pathways depending on the reaction system. When substrate is the primary reactant, the reaction usually involves an electron transfer or hydrogen abstraction process. This typically results in a highly reactive free radical form of the substrate which can react with molecular oxygen; the reduced sensitizer is regenerated, often by reaction with oxygen. Photooxidations of this type, which proceed by way of free radical intermediates, are often termed Type I processes (6).

When oxygen is the primary reactant, energy transfer from triplet dye to ground state oxygen may occur, raising the oxygen molecule to an excited state, probably the $^1\Delta_g$ state, as first suggested by Kautsky et al. (7; also see 3). Alternately, a labile, energy-rich addition compound of the triplet dye and oxygen may be formed (see 8). Both singlet oxygen and the oxygen-sensitizer addition compound are very reactive and can oxidize a variety of susceptible substrates. Photooxygenations of this type, which involve the participation of only electronically-excited states, are termed Type II processes (8). In photodynamic systems with biological macromolecules, both Type I and Type II processes appear to be involved, with the relative participation depending on the particular system.

THE SENSITIZED PHOTOOXIDATION OF CARBOHYDRATES

Investigations have been made of the dye-sensitized photooxidation of only a very few kinds of polysaccharides, in particular cellulose (as exemplified by cotton fibers) and certain of its derivatives, and hyaluronic acid and related compounds. Cellulose is a linear polymer of glucose subunits linked by C-1 to C-4 bonds; recent studies suggest that a native cellulose molecule consists of approximately 14,000 glucose units, giving a molecular weight greater than 2×10^6. Many dyes sensitize the "phototendering" of cotton and other textile fibers, a degradative process which consumes oxygen and which results in a decrease in the mechanical strength of the fiber (9). This phenomenon has been studied extensively because of its obvious practical importance to the textile industry (9). Most of this research has involved dyes used commercially for dyeing cotton, in particular the anthraquinone vat dyes. The classes of dyes usually employed in investigations of photodynamic action, such as the acridines, thiazines, xanthenes, etc., have not been studied to any great extent with carbohydrates.

Several different reaction mechanisms appear to be involved in the dye-sensitized phototendering of cellulose (10-12). Some experiments demonstrate clearly that illumination of dyed cellulose fibers, especially in the presence of moisture, can produce a volatile oxidizing agent which diffuses to near-by undyed fibers to produce tendering (13). The oxidizing entity has been variously postulated to be hydrogen peroxide (13,14), semireduced oxygen (O_2^-) (15), or singlet oxygen (13). Little is known of the chemical changes produced in the cellulose molecule by the diffusible oxidizing agent. With the present interest in the role of singlet oxygen in dye-sensitized photooxygenation reactions, and with the presently available techniques for producing and identifying singlet oxygen (3), it would be of interest to re-examine this whole problem.

The other generally proposed mechanism for the phototendering of cellulose is based on studies of the sensitized photooxidation of alcohols as model compounds, in particular with anthraquinones as sensitizers (10,12). In such reactions, the quinone form of the light-excited dye abstracts a hydrogen from the α-carbon of the alcohol to produce an alcohol radical. The dye is simultaneously converted to a semiquinone which reacts rapidly with molecular oxygen to regenerate the quinone form. Radicals of primary alcohols react with oxygen yielding the corresponding aldehyde or carboxylic acid, while radicals of secondary alcohols give the corresponding ketone (16,17). Similar results are obtained with hexitols, such as sorbitol, as model compounds for the glucose residues of cellulose; the initial oxidation product with anthraquinones is the corresponding hexose resulting from oxidation of the primary alcohol group (18). Electron spin resonance measurements demonstrate the production of free radicals from anthraquinones on illumination in photodynamic systems (19). Further, most tendering anthraquinone vat dyes initiate the polymerization of methyl methacrylate by free radical production on illumination (20). Thus the role of free radicals in anthraquinone-sensitized photooxidations seems well established.

In summary, it appears that the dye-sensitized phototendering of cellulose can proceed by two pathways; the relative involvement of the two paths depends on the sensitizing dye, the way in which

the dye is bound to the fiber, the humidity, and the concentration
of oxygen in the immediate environment (12).

Hyaluronic acid, a mucopolysaccharide, is a linear polymer of
alternating D-glucuronic acid and N-acetyl-D-glucosamine units with
molecular weights ranging up to several million. It is the major
component of the jelly-like ground substance of animal tissues and
also occurs in the vitreous humor of the mammalian eye. Solutions
of hyaluronic acid are depolymerized, as evidenced by a decrease
in viscosity, on illumination in the presence of molecular oxygen
and sensitizing dyes such as hematoporphyrin (21), phylloerythrin
(22), hypericin (22), acriflavine (23), acridine orange (23), crystal violet (23), methylene blue (23), neutral red, (23), and riboflavine (23,24). The chemistry of the depolymerization of hyaluronic acid on photodynamic treatment is not known. It may be that
the reaction pathway is somewhat different from the mechanisms
usually suggested for photodynamic action, since the reaction is
accelerated by reducing agents such as ascorbic acid and perhaps by
trace amounts of heavy metals. It has been suggested that free radicals, formed by illumination of the dye-reducing agent part of the
systems, bring about depolymerization of the hyaluronic acid (23).
The suggested role of free radicals is supported by the observation
that cationic dye-hyaluronic acid or cationic dye-heparin (also a
mucopolysaccharide) complexes become paramagnetic on illumination
(25).

SENSITIZED PHOTOOXIDATION OF NUCLEIC ACIDS AND OTHER POLYNUCLEOTIDES

Nucleic acids, with molecular weights up to 10^8 or more, are
linear biopolymers composed of nucleotide units, each of which is
made up of a phosphate, a pentose, and a heterocyclic organic base
(purine, pyrimidine). Nucleotides are joined together by a bond
between the 3' carbon of the pentose of one nucleotide and the
phosphate group of the next nucleotide in the chain; thus the "backbone" of the polynucleotide is made up of a -sugar-phosphate-sugar-
phosphate-, etc., sequence. In the three-dimensional model, nucleic
acids exist as a helix composed of two polynucleotide chains held
together by base to base hydrogen bonds. Although many millions of
different kinds of nucleic acids must occur in living material,
they are composed of only a few types of purines (adenine, guanine),
pyrimidines (thymine, cytosine, uracil) and pentoses (ribose, 2-
deoxyribose). Self replicating structures, such as chromosomes,
contain deoxyribonucleic acid (DNA) characterized by 2-deoxyribose
as the pentose and thymine and cytosine as the pyrimidines. The
other type of nucleic acid, ribonucleic acid (RNA), differs from
DNA in that it contains ribose as the sugar and uracil rather than
thymine.

It was observed a few years ago that guanine residues were

destroyed selectively during the methylene blue-sensitized photodynamic treatment of DNA (26). This same selectivity was also observed with the free bases, nucleosides and nucleotides from nucleic acids. For example, deoxyguanylic acid is rapidly photooxidized with methylene blue, thymidylic acid is oxidized only
slowly, while deoxyadenylic, deoxycytidylic and uridylic acids are
not oxidized appreciably (26). In general, di- and tri-substituted
purines are rapidly photooxidized, while purine itself, and mono-
substituted purines are resistant (27; also see 28,29). The 8-aza
derivatives of purines are resistant to photooxidation which suggests that susceptible derivatives must possess a structure capable
of photooxidation at the 8-position (30). Zenda, et al. (31), on
the basis of their examination of a large number of compounds, concluded that a base must have an imidazole ring and a lactim structure involving nitrogens 1 and 3 in the pyrimidine part of the
purine ring in order to be susceptible to photooxidation.

     A variety of dyes, including eosin Y, fluorescein, neutral
red, rose bengal, thionine and toluidine blue sensitize the photooxidation of guanine derivatives, whereas alizarin red, crystal
violet, methyl orange, proflavine, pyronine and safranine 0 do not
sensitize (27). With most dyes, the rate of photooxidation of the
nucleic acid bases increases at high pH, indicating that the anionic
form of the base is the most susceptible to photooxidation (27).
Although several studies have been made, little is known of the
primary reaction products resulting from the photooxidation of
guanine and its derivatives. Both ring systems of guanosine are
broken on illumination in the presence of methylene blue; the main
products identified are guanidine, ribose, ribosylurea and urea
(32,33). With lumiflavin as sensitizer, carbon dioxide, guanidine,
parabanic acid, and a variety of unidentified products are formed
from guanine (see 34). Laser and tungsten lamp illumination of
guanine in the presence of methylene blue gave formylguanidine,
parabanic acid and urea (35). Studies on the methylene blue-sensitized photooxidation of C-14 labelled uric acid indicate that
urea may be a primary product from the C-2 part of the molecule,
while urea from the C-8 part of the molecule may be a secondary
product (30). Mechanistic studies of the photooxidation of purine
derivatives with rose bengal suggest the involvement of peroxide
intermediates (36).

     Gaffron (37) first showed that nucleic acids could be photo-
oxidized with dyes such as methylene blue and erythrosin. Since
then, a large number of studies have been made on the photodynamic
treatment of nucleic acids and synthetic polynucleotides. Water
is apparently required for the photodynamic inactivation of RNA
(38). The illumination of dye-DNA mixtures leads to the formation
of free radicals; there is a good correlation with the free radical
concentrations produced with a given dye and the rate of photoinactivation of the DNA (39). The photodynamic treatment of nucleic

acids often alters their biological properties. For example, the
DNA transforming principle from some bacteria is destroyed by photodynamic treatment (40). The ability of tobacco mosaic virus RNA
to infect tobacco plants (41) and to stimulate the incorporation of
certain amino acids into protein (42) is also lost. Transfer-RNA
loses activity on photodynamic treatment (43,44). The synthetic
polynucleotide, poly-UG (a copolymer of uridylic acid and guanylic
acid) loses its "messenger" activity, that is, its capability of
directing the incorporation of amino acids into proteins in biological systems (44,45).

In many cases, photooxidation alters the physicochemical
properties of nucleic acids. For example, prolonged photodynamic
treatment causes a decrease in the viscosity of nucleic acid solutions (46). This reaction is very slow with DNA since double-chain
scission of the double helix, which would be necessary to produce
a decrease in viscosity, occurs only infrequently (47,48). Photodynamic treatment does make DNA much more susceptible to subsequent
chain breakage by hot perchloric acid (47) and to enzymatic degradation (49). It also changes the apparent solubility of DNA in
bacterial cells by causing a crosslinking of the DNA to bacterial
protein (50). DNA becomes covalently linked to 3,4-benzpyrene on
illumination (51). Photodynamically-inactivated tobacco mosaic
virus RNA (52) and T_4 phage DNA (26) as well as photodynamically-
treated poly-UG (45) show essentially no decrease in their ultra-
centrifuge sedimentation behavior. This indicates that there is
little rupture of the sugar-phosphate backbone with the illumination times necessary for the loss of biological activity. The
"melting" temperature of DNA decreases on photodynamic treatment
suggesting a change in the stability of the secondary structure
(46); with DNA's from different organisms, the effect is proportional to the mole fraction of guanine residues in the DNA (46,53).

THE SENSITIZED PHOTOOXIDATION OF PROTEINS

As indicated in the introduction, the dye-sensitized photooxidation of many different proteins has been studied (see listings
in reference 1). Proteins are polymers composed of amino acids
linked together by peptide bonds, i.e., a bond formed by the con-
densation of the amino group of one amino acid with the carboxyl
group of another. Proteins have chain lengths in the range of 100-
1,000 or more amino acid residues, and therefore molecular weights
of approximately 10^4 to over 10^6. Typical proteins are composed of
various sequences of the same twenty-odd amino acids. In most
proteins at least part of the polypeptide chain is coiled into a
helix; typically the chain is folded further in a complex fashion
giving the protein molecule a characteristic three-dimensional conformation maintained by electrostatic interactions, hydrogen bonds
and disulfide bonds. Thus some of the amino acid residues in a

protein will be located in a peripheral "exposed" position, while
others will be "buried" within the molecule. The location of an
amino acid residue in the three-dimensional structure of the protein can be important in determining its sensitivity to photooxidation as discussed below. Irradiation of proteins with short wavelength ultraviolet light or with ionizing radiation leads to the
rupture of peptide and disulfide bonds. In contrast, with dye-
sensitized photooxidation, there is no breakage of these bonds
(see 54,55); all damage to the protein molecule results from the
alteration of certain amino acid side chains, including those of
cysteine, histidine, methionine, tryptophan and tyrosine. Thus an
understanding of the photochemistry of these five amino acids and
their derivatives is important for elucidating the mechanisms of
the photodynamic alteration of proteins.

Sensitized Photooxidation of Free Amino Acids and Their Derivatives

     Solutions of tryptophan are photooxidized with certain dyes,
such as proflavine, to yield two classes of products, kynurenines
and melanines (56). Tryptophan is converted to N-formyl kynurenine,
while the tryptophan model compound, 3-methyl indole (skatole), is
converted to o-acetyl-formanilide on photooxidation with methylene
blue (57). When the photooxidation of tryptophan is carried out in
acetic acid solution, the solvent is incorporated into the reaction
products to give β-carboline derivatives (58). The photooxidation
of tyrosine with methylene blue involves rupture of the ring and
carbon dioxide formation, however, little is known of the organic
intermediates involved (59). A model compound for histidine, 4-
methylimidazole, gives rise to acetylurea on photooxidation with
methylene blue (57), while a number of unidentified organic products
result from histidine photooxidation (57,60). Cysteine is quanti-
tatively photooxidized to cysteic acid with crystal violet (61).

     The photooxidation products of at least one amino acid, meth-
ionine, depend on the sensitizing dye and on the reaction conditions
used. Methionine sulfoxide is the sole oxidation product obtained
with dyes such as methylene blue (59), proflavine (56), rose bengal
(62), and with a variety of other dyes (unpublished results). The
predominant reaction pathway with these dyes is probably that involving singlet oxygen. With methylene blue at pH 8, one mole of
oxygen is consumed per mole of methionine photooxidized, to produce
one mole of methionine sulfoxide and presumably one mole of hydrogen peroxide (59). At low pH, however, we find that one-half mole
of oxygen is consumed. With flavine dyes, only a small fraction of
the methionine photooxidized is converted to the sulfoxide (63),
the rest appearing as methional (64). This latter product apparently results from a photochemical electron-abstraction process (65).

     In some cases, by the proper selection of the sensitizing dye and of the reaction conditions, certain amino acids can be selectively photooxidized in the presence of others. With most dyes, for example, histidine and tyrosine are not photooxidized at low pH, whereas the other three susceptible amino acids are rapidly oxidized (66,67). Furthermore, only methionine and tryptophan are photooxidized in acetic and formic acids with proflavine (56). At low temperatures and in acetic acid solutions, only methionine is photooxidized at an appreciable rate with rose bengal or methylene blue (62). Cysteine and methionine are selectively photooxidized with dyes such as bromthymol blue, congo red and methyl orange (61); under some conditions only cysteine is photooxidized with crystal violet and with cresol red (61,67). The incorporation of an amino acid into small peptides and proteins can alter its photodynamic behavior. For example, blocking the α-amino group of tryptophan, e.g., by incorporating it into a peptide, makes it insensitive to photooxidation with hematoporphyrin (68). We have found (unpublished work) that the incorporation of methionine into di- and tri-peptides also alters the quantum yield of its photooxidation with eosin Y and lumiflavin.

Biological Changes in Proteins During Photooxidation

Proteins play a variety of physical and chemical roles in the living organism. In general, photooxidation of proteins alters their properties to the point where they can no longer serve their usual purpose. The effects of photodynamic action on many different kinds of proteins has been studied (see listing in reference 1). Two protein hormones, insulin (69) and angiotensinamide (70), are inactivated on photooxidation. Certain snake venoms (71) and bacterial toxins (72) can be detoxified on illumination in the presence of dyes; further, the characteristic biological activity of antigens (73) and antibodies (74) can be destroyed. Studies have been made of the photodynamic sensitivity of a wide variety of enzymes, including representatives of all categories (oxidoreductases, transferases, hydrolases, lyases, isomerases and ligases), as tabulated in reference (1). Of those enzymes studied, only very few are resistant to photodynamic inactivation (1,73).

Physicochemical Changes in Proteins During Photooxidation

A large number of kinetic studies have been made of the dye-sensitized photooxidation of proteins (1,73). Proteins are typically photooxidized via reactions which are first order with respect to protein concentration (1,73). With most dyes and proteins, the rates of photooxidation are low under acidic conditions, increase with increasing pH around neutrality, and usually level off at high pH (1,73,75,76). At low temperatures, experimental activation energies are usually quite small, in the range of a few kcal/mole (75). The quantum yields for protein photoinactivation typically increase with increasing oxygen concentration up to about 20% oxygen (77). Yields also increase with increasing dye concentration, go through a maximum, and then decline in the range of 10^{-3} M (75). Many chemically different dyes have been studied in order to establish relationships between molecular structure and sensitizing efficiency. Dyes of certain types (azo,indophenol, nitro, oxazine, thiazole, etc.) are not active in solution, whereas most of the acridine, anthraquinone, azine, flavine, thiazine and xanthene dyes as well as a number of porphyrin derivatives are good sensitizers (1,55,73).

Many proteins show marked physicochemical changes as a result of dye-sensitized photooxidation. Fibrous proteins such as collagen (78), silk and wool (79,80), for example, typically show changes in their mechanical properties upon photodynamic treatment. Many proteins show a decrease in absorption at 2800 Å as a result of the photodynamic destruction of the aromatic side chains (81). Also, photooxidation of some proteins leads to increased heat sensitivity (82), increased sensitivity to digestion by proteolytic enzymes (82,83), as well as changes in surface tension (84), solubility (85), viscosity (86), electrophoretic mobility (87), conformation (82), light scattering (88), and changes in the diffusion (89), sedimentation (89), polarographic (90), optical rotation (91), and metal-binding (83) behavior.

Selective Effects in the Photooxidation of Proteins

Since, as discussed above, the photooxidizable amino acids react at different rates depending on the reaction conditions, certain selective effects in the photooxidation of proteins can be obtained by the choice of the sensitizing dye and by using various reaction conditions. These selective effects, i.e., the photooxidation of one type of amino acid residue or of a selected set of residues in the protein, result from: a) the degree of "accessibility" of the various residues in the protein; b) the ionic state of the residues due to conditions such as pH; c) the use of a sensitizer selective for various types of amino acid residues; or d) some sort of specific association between the sensitizer and the protein molecule. These four factors, which can be used in combination with one another for obtaining various degrees of selectivity in the photooxidation of proteins, are discussed more fully below.

a). Due to the three-dimensional folding of the protein molecule, certain of the photooxidizable amino acid residues are located at the surface of the molecule and will be photooxidized at a faster rate than those residues which are variously "inaccessible". Ray and Koshland (92) were the first to use this phenomenon

in a kinetic approach for deducing the number of accessible and
inaccessible photooxidizable residues in a protein. In the methylene blue-sensitized photooxidation of the enzyme phosphoglucomu-
tase, it was possible to distinguish accessible and inaccessible
classes of methionyl and histidyl residues, and to deduce the number of residues in these classes from the rates of photooxidation
of the free amino acids in solution, from the rate of the photodynamic inactivation of the enzyme, and from semilogarithmic plots
of the photooxidation of each type of residue.

Other examples involving this type of selective effect in the
photooxidation of proteins include the studies of Weil, et al. (81)
who photooxidized the histidyl residues in insulin by illumination
with methylene blue at low temperatures. Upon unfolding the protein with 8 M urea or by raising the temperature prior to the illumination,they were able to obtain the additional photooxidation
of tyrosyl residues. By removing the 20-residue S-peptide from
the enzyme ribonuclease-S (RNase-S), Kenkare and Richards (93) investigated the accessibility of the four histidyl residues in RNase-
S to methylene blue-sensitized photooxidation. Jori, et al. (94)
studied the accessibility of methionyl residues in ribonuclease A
(RNase A) to photooxidation sensitized by hematoporphyrin by progressively unfolding the enzyme with increasing concentrations of
acetic acid. Using this technique, it was possible to demonstrate
three degrees of exposure of methionyl (Met) residues; Met-79 and
Met-30 were the most inaccessible, Met-13 was somewhat inaccessible,
and Met-29 was largely accessible.

b). Some degree of selectivity in the photooxidation of amino
acid residues in proteins can be obtained by controlling the ionic
state of the amino acid side chains. For example, since histidyl
residues in proteins are susceptible to photooxidation only when
the imidazole moiety is uncharged, these residues are not photooxidized to any great extent at low pH ranges (1,95). In addition,
the pH-dependence of the dye-sensitized photoinactivation of enzymes
is often the same as that of the photooxidation of histidine, especially when histidyl residues are the only ones altered in the enzyme (96,98). It is also useful to exclude histidyl photooxidation
by working under acidic conditions in order to obtain the selective
destruction of other types of residues, such as tryptophyl (99) or
methionyl residues (62,100).

c). Many dyes, especially when used with particular reaction
conditions such as solvent system and pH, can be very specific for
the photooxidation of a given type of amino acid residue. One of
the best examples of the high degree of selectivity which can be
obtained with certain sensitizers is the cresol red- or crystal
violet-sensitized photooxidation of cysteinyl residues. No detectable photooxidation of tryptophyl, tyrosyl, histidyl, or methionyl residues was observed with these dyes in aqueous systems buf-

fered from pH 2.5 to pH 9, or in 5% to 95% aqueous solutions of acetic acid (61). With reduced ribonuclease (101) or lysozyme (61),
which contain free sulfhydryl groups, illumination with cresol red
or crystal violet results in the photooxidation of the cysteinyl
residues to produce the corresponding sulfonated enzymes; no photooxidation is observed with the native enzymes, which lack free
sulfhydryl groups.

Hematoporphyrin sensitizes the selective photooxidation of
methionyl residues over a wide pH range below pH 6.5 and in aqueous
acetic acid solutions (68). In addition, by progressively unfolding lysozyme (102) and RNase A (94) with acetic acid, it is possible
to obtain the selective photooxidation of certain methionyl residues.
Methionyl residues are also altered selectively by using methylene
blue or rose bengal in acetic or formic acid solutions; the only
other type of residue altered under these conditions is tryptophan,
and that only very slowly (62). Using this latter technique, the
selective photooxidation of methionyl residues to methionine sulfoxide in lysozyme (100) and RNase A (62) was obtained.

Methyl orange, congo red, bromthymol blue, and 4,4'bis (di-
methylamino)thiobenzophenone sensitize the selective photooxidation
of methionyl and cysteinyl residues (61). Thus it should be possible
to use these sensitizers for the selective modification of methionyl residues in non-sulfhydryl proteins; or, if methionine sulfoxide is produced with these dyes, it should be possible to obtain
the net oxidation of only cysteinyl residues by subsequently reducing the methionine sulfoxide to methionine by chemical treatment
as described below.

Methionyl and tryptophyl residues in proteins are selectively
photooxidized by illumination with proflavine in 98% to 100% formic
or acetic acid. The resulting methionine sulfoxide residues can
be converted to methionyl residues by subsequent treatment with
mercaptoethanol or thioglycolic acid. Using this procedure,
Galiazzo, et al. (99) obtained the net selective photooxidation of
tryptophyl residues in lysozyme.

Thus rather highly selective techniques have been developed
for the photooxidation of cysteinyl, methionyl, and tryptophyl
residues in proteins by the use of certain sensitizers. In addition, there is evidence for some degree of selectivity for the
photooxidation of tyrosyl residues in proteins by using flavine
mononucleotide (FMN) as a sensitizer (103). Also, a number of
investigators have obtained the selective photooxidation of histidyl
residues in certain proteins because of the slower reactivity, or
the lesser degree of accessibility of the other types of susceptible
residues present (96,98,104).

Many sensitizers, while not highly specific, still exhibit

some degree of selectivity in the photooxidation of residues in certain proteins. For example, Yamagata, et al. (105) found that histidyl and tryptophyl residues were altered in the methylene blue-sensitized photoinactivation of RNase T$_1$, whereas only histidyl residues were altered with riboflavine. Likewise, della Pietra and Dose (106) found that in the photoinactivation of yeast alcohol dehydrogenase, there was an apparent destruction of only cysteinyl residues with hematoporphyrin as a sensitizer, and cysteinyl, tyrosyl, tryptophyl and/or histidyl residues with FMN.

d). When the sensitizer is in close association with the protein, it is possible to obtain the highly selective photooxidation of only those few susceptible amino acid residues located in close proximity to one another and to the sensitizer due to the tertiary structure of the protein. This technique should be very useful for the investigation of the three-dimensional structure of proteins in solution under a variety of conditions (101,107).

Pyridoxal 5'-phosphate, when bound to the lysyl residue at the active site of 6-phosphogluconate dehydrogenase, sensitizes the photooxidation of two of the 12 histidyl residues and three of the 8 cysteinyl residues present in the molecule (108). Thus these altered residues were thought to be located near the active site of the enzyme. Likewise, Sawada (109) found that complexing of the chromophoric substrate analogue 4-thiouridylic acid with RNase was necessary for the sensitized photoinactivation of the enzyme. In this case, the substrate analogue sensitized the photooxidation of only one tyrosyl residue. Breslow, et al. (110) found that one or two histidyl residues were preferentially photooxidized when the 1:1 complex of protoporphyrin IX and sperm whale apomyoglobin was illuminated, causing the dissociation of the complex. Thus it was concluded that these residues were located closest to the binding site for the protoporphyrin IX.

In addition to these weaker types of associations discussed above, sensitizing moieties can also be covalently linked to known positions in a protein molecule where they can sensitize the photooxidation of near-by susceptible residues. Illumination of 41-DNP-RNase A, i.e., RNase A to which a dinitrophenyl chromophore has been covalently linked at the lysine-41 position of the active site, caused the alteration of one histidyl, one methionyl, and one tyrosyl residue (101). Thus these residues are presumably located near the active site of the enzyme. By contrast, illumination of denatured 41-DNP-RNase A in which the chromophore, although still covalently attached, was free to contact any part of the molecule, resulted in the photooxidation of all the susceptible amino acid residues in the enzyme.

A sensitizing moiety occurs naturally in some proteins, such as the cytochromes. Jori, et al. (111) found that one methionyl residue (Met-80) was photooxidized upon illumination of horse heart cytochrome c at pH 5.9; at pH 8.2, a histidyl residue (His-18) was altered in addition to Met-80. Additional histidyl, methionyl, tryptophyl, and tyrosyl residues were photooxidized when cytochrome c was illuminated with hematoporphyrin as added sensitizer. The presumably photodynamic destruction of cytochromes a₃, b, and c was observed by Epel and Butler (112); this photodestruction is probably sensitized by the natural chromophores of the cytochromes as in the case of horse heart cytochrome c.

As described above, it is clear that through an association of the sensitizer with the protein, one can obtain the photooxidation of a small number of susceptible amino acid residues located in close proximity to one another and to the sensitizing moiety as a result of the tertiary structure of the protein. Just how near the sensitizer a susceptible residue would need to be located in order to be photooxidized depends upon the reaction mechanism, and especially upon the diffusability of the oxidizing species. In any event, the susceptible residues located closest to the chromophore would be expected to be photooxidized with the greatest probability, and therefore this type of selectivity in the photooxidation of proteins can be used for exploring various regions of protein molecules.

REFERENCES

1. J. D. Spikes and R. Livingston, Adv. Radiation Biol., 3, 29 (1969).
2. H. von Tappeiner, Ber. Deut. Chem. Ges., 36, 3035 (1903).
3. C. S. Foote, Science, 162, 963 (1968).
4. J. S. Bellin, Photochem. Photobiol., 8, 383 (1968).
5. N. K. Bridge and G. Porter, Proc. Roy. Soc. (London), A244, 259 (1958).
6. K. Gollnick, Adv. Photochem., 6, 1 (1968).
7. H. Kautsky, H. de Bruijn, R. Neuwirth and W. Baumeister, Chem. Ber., 66, 1588 (1933).
8. K. Gollnick and G. O. Schenck, Pure Appl. Chem., 9, 507 (1964).
9. "Photochemistry In Relation To Textiles", The Society of Dyers and Colourists, Bradford, Yorkshire, (1950). Proceedings of a Symposium held at Harrogate, England, 22-24 September 1949.
10. J. Bourdon and B. Schnuriger, pp. 60-131 in "Physics And Chemistry Of The Organic Solid State", Vol. III., edited by D. Fox, M. M. Labes and A. Weissberger, Interscience, New York, (1967).
11. G. O. Phillips, Adv. Carbohydrate Chem., 18, 9 (1963).
12. G. O. Phillips and J. C. Arthur, Jr., Textile Res. J., 34, 497, 572 (1964).
13. G. S. Egerton, J. Soc. Dyers Colourists, 65, 764 (1949).
14. C. H. Bamford and M. J. S. Dewar, J. Soc. Dyers Colourists, 65, 674 (1949).

15. G. S. Egerton, Nature, 204, 1153 (1964).
16. J. L. Bolland and H. R. Cooper, Proc. Roy. Soc. (London)
    A225, 405 (1954).
17. C. F. Wells, Trans. Faraday Soc., 57, 1703, 1719 (1961).
18. G. O. Phillips, P. Barber and T. Rickards, J. Chem. Soc.,
    3443 (1964).
19. P. J. Baugh, G. O. Phillips and J. C. Arthur, Jr., J. Phys.
    Chem., 70, 3061 (1966).
20. J. J. Moran and H. I. Stonehill, J. Chem. Soc., 788 (1957).
21. A. Castellani, Giorn. Biochim., 3, 19 (1954).
22. A. Castellani and V. Torlone, J. Pathol. Bacteriol., 72, 505
    (1956).
23. L. Sundblad and E. A. Balazs, pp. 229-250 in "The Amino Sugars",
    Vol. II B, edited by R. W. Jeanloz and E. A. Balazs, Academic
    Press, New York (1966).
24. G. Matsumura, A. Herp and W. Pigman, Rad. Res., 28, 735 (1966).
25. E. A. Balazs, G. O. Phillips and M. D. Young, Biochim. Biophys.
    Acta, 141, 382 (1967).
26. M. I. Simon and H. Van Vunakis, J. Mol. Biol., 4, 488 (1962).
27. M. I. Simon and H. Van Vunakis, Arch. Biochem. Biophys., 105,
    197 (1964).
28. A. Wacker, G. Türck and A. Gerstenberger, Naturwiss., 50, 377
    (1963).
29. E. R. Lochmann, W. Stein and K. Haefner, Z. Naturforsch.,
    19b, 838 (1964).
30. P. A. Friedman, Biochim. Biophys. Acta, 166, 1 (1968).
31. K. Zenda, N. Saneyoshi and G. Chihara, Chem. Pharm. Bull.
    (Tokyo), 13, 1108 (1965).
32. K. S. Sastry and M. P. Gordon, Biochim. Biophys. Acta, 129,
    42 (1966).
33. L. A. Waskell, K. S. Sastry and M. P. Gordon, Biochim. Biophys.
    Acta, 129, 49 (1966).
34. J. S. Sussenbach and W. Berends, Biochim. Biophys. Acta, 95,
    184 (1965).
35. Y. LeRoux, C. Nofre and G. Peres, C. R. Acad. Sci. (Paris),
    Ser. D, 266, 1323 (1968).
36. T. Matsuura and I. Saito, Tetrahedron, 24, 6609 (1968).
37. H. Gaffron, Biochem. Z., 179, 157 (1926).
38. T. Ito and J. Amagasa, Currents Mod. Biol., 1, 329 (1968).
39. M. Delmelle and J. Duchesne, Studia Biophysica, 3, 121 (1967).
40. J. S. Bellin and L. I. Grossman, Photochem. Photobiol., 4,
    45 (1965).
41. M. Chessin, Science, 132, 1840 (1960).
42. B. Singer and H. Fraenkel-Conrat, Biochem., 5, 2446 (1966).
43. A. Tsugita, Y. Okada and K. Uehara, Biochim. Biophys. Acta,
    103, 360 (1965).
44. P. Chandra and A. Wacker, Z. Naturforsch., 21b, 663 (1966).
45. M. I. Simon, L. Grossman and H. Van Vunakis, J. Mol. Biol.,
    12, 50 (1965).

46. D. Freifelder, P. F. Davison and E. P. Geiduschek, Biophys.
    J., 1, 389 (1961).
47. J. S. Bellin and C. A. Yankus, Biochim. Biophys. Acta, 112,
    363 (1966).
48. M. Kuwano, Y. Hayashi, H. Hayashi and K. Miura, J. Mol. Biol.,
    32, 659 (1968).
49. H. Dellweg and W. Oprée, Biophysik, 3, 241 (1966).
50. K. C. Smith, Biochem. Biophys. Res. Commun., 8, 157 (1962).
51. G. Reske and J. Stauff, Z. Naturforsch., 20b, 15 (1965).
52. K. S. Sastry and M. P. Gordon, Biochim. Biophys. Acta, 129,
    32 (1966).
53. J. S. Bellin and L. I. Grossman, Photochem. Photobiol., 4,
    45 (1965).
54. C. A. Ghiron and J. D. Spikes, Photochem. Photobiol., 4, 13
    (1965).
55. J. D. Spikes and B. W. Glad, Photochem. Photobiol., 3, 471
    (1964).
56. C. A. Benassi, E. Scoffone, G. Galiazzo and G. Jori, Photochem.
    Photobiol., 6, 857 (1967).
57. G. Rhodes and P. D. Gardner, (unpublished results).
58. G. Jori, G. Galiazzo and G. Gennari, Photochem. Photobiol.,
    9, 179 (1969).
59. L. Weil, W. G. Gordon and A. R. Buchert, Arch. Biochem.
    Biophys., 33, 90 (1951).
60. S. Gurnani and M. Arifuddin, Photochem. Photobiol., 5, 341
    (1966).
61. G. Jori, G. Galiazzo and E. Scoffone, Int. Jour. Protein Res.,
    1, 289 (1969).
62. G. Jori, G. Galiazzo, A. Marzotto and E. Scoffone, Biochim.
    Biophys. Acta, 154, 1 (1968).
63. J. D. Spikes, Biophys. J., 9, A-272 (1969).
64. K. Enns and W. H. Burgess, J. Am. Chem. Soc., 87, 5766 (1965).
65. S. F. Yang, H. S. Ku and H. K. Pratt, J. Biochem., 242, 5274
    (1967).
66. L. A. Ae. Sluyterman, Biochim. Biophys. Acta, 60, 557 (1962).
67. J. S. Bellin and C. A. Yankus, Arch. Biochem. Biophys.,
    123, 18 (1968).
68. G. Jori, G. Galiazzo and E. Scoffone, Biochem., 8, 2868 (1969).
69. G. Weitzel, W. Schaeg, G. Boden and B. Willms, Z. Naturforsch.,
    20^b, 497 (1965).
70. A. C. M. Paiva and T. B. Paiva, Biochim. Biophys. Acta, 48,
    412 (1961).
71. W. Kocholaty, Toxicon, 3, 175 (1966).
72. D. A. Boroff and B. R. DasGupta, J. Biol. Chem., 239, 3694
    (1964).
73. Z. Vodrážka, Chem. Listy, 53, 829 (1959).
74. A. Tyler, J. Immunol., 51, 329 (1945).
75. B. W. Glad and J. D. Spikes, Radiation Res., 27, 237 (1966).
76. M. L. MacKnight and J. D. Spikes, Experientia, in press (1970).

77. C. F. Hodgson, E. B. McVey and J. D. Spikes, Experientia, 25, 1021 (1969).
78. J. D. Spikes and C. A. Ghiron, pp. 309-336 in "Physical Processes in Radiation Biology", edited by L. G. Augenstein, R. Mason and B. Rosenberg, Academic Press, New York (1964).
79. G. S. Egerton and K. M. Shah, Nature, 202, 81 (1964).
80. I. H. Leaver and G. C. Ramsay, Photochem. Photobiol., 9, 531 (1969).
81. L. Weil, T. S. Seibles and T. T. Herskovits, Arch. Biochem. Biophys., 111, 308 (1965).
82. T. R. Hopkins and J. D. Spikes, Photochem. Photobiol., in press (1970).
83. P. Friedrich, L. Polgar and G. Szabolcsi, Acta Physiol. Acad. Sci. Hung., 25, 217 (1964).
84. L. Santamaria, R. Santamaria and M. Ciarfuglia, Atti. Soc. Ital. Patol., 5, 457 (1957).
85. J. S. Bellin and G. Entner, Photochem. Photobiol., 5, 251 (1966).
86. R. Santamaria, G. G. Corigliano and L. Santamaria, Boll. Soc. Ital. Biol. Sper., 40, 1801 (1964).
87. Z. Vodrážka, J. Čejka and J. Salák, Biochim. Biophys. Acta, 52, 342 (1961).
88. L. Weil and A. R. Buchert, Arch. Biochem. Biophys., 34, 1 (1951).
89. J. M. Brake and F. Wold, Biochim. Biophys. Acta, 40, 171 (1960).
90. S. Fiala, Biochem. Z., 320, 10 (1949).
91. K. Sekiya, S. Mii and Y. Tonomura, J. Biochem. (Tokyo), 57, 192 (1965).
92. W. J. Ray, Jr. and D. E. Koshland, Jr., J. Biol. Chem., 237, 2493 (1962).
93. U. W. Kenkare and F. M. Richards, J. Biol. Chem., 241, 3197 (1966).
94. G. Jori, G. Galiazzo, A. M. Tamburro and E. Scoffone, J. Biol. Chem., in press (1970).
95. B. R. DasGupta and D. A. Boroff, Biochim. Biophys. Acta, 97, 157 (1965).
96. K. A. Freude, Biochim. Biophys. Acta, 167, 485 (1968).
97. E. W. Westhead, Biochem., 4, 2139 (1965).
98. N. E. Vorotnitskaya, G. F. Lutovinova and O. L. Polyanovsky, pp. 131-142 in "Pyridoxal Catalysis: Enzymes and Model Systems", edited by E. E. Snell, A. E. Braunstein, E. S. Severin and Yu. M. Torchinsky, Interscience Publishers, New York (1968).
99. G. Galiazzo, G. Jori and E. Scoffone, Biochem. Biophys. Res. Commun., 31, 158 (1968).
100. G. Jori, G. Galiazzo, A. Marzotto and E. Scoffone, J. Biol. Chem., 243, 4272 (1968).
101. E. Scoffone, G. Galiazzo and G. Jori, Biochem. Biophys. Res. Commun. 38, 16 (1970).

102. G. Galiazzo, A. M. Tamburro and G. Jori, Eur. J. Biochem., 12, 362 (1970).
103. M.L. MacKnight and J. D. Spikes, Boll. Chim. Farm., in press (1970).
104. Y. Ichikawa and T. Yamano, Tokushima J. Exptl. Med., 10, 156 (1963).
105. S. Yamagata, K. Takahashi and F. Egami, J. Biochem. (Tokyo), 52, 261 (1962).
106. D. della Pietra and K. Dose, Biophysik, 2, 347 (1965).
107. G. Galiazzo, G. Jori and E. Scoffone, In press, in, "Research Progress in Organic-Biological and Medicinal Chemistry", Vol. III, edited by U. Gallo and L. Santamaria, North Holland Publishing Co., Amsterdam (1970).
108. M. Rippa and S. Pontremoli, Arch. Biochem. Biophys., 113, 112 (1969).
109. F. Sawada, J. Biochem. (Tokyo), 65, 767 (1969).
110. E. Breslow, R. Koehler and A. W. Girotti, J. Biol. Chem., 242, 4149 (1967).
111. G. Jori, G. Gennari, G. Galiazzo and E. Scoffone, FEBS Letters, 6, 267 (1970).
112. B. Epel and W. L. Butler, Science, 166, 621 (1969).

PHOTOLYTIC OXIDATION OF ISOTACTIC POLYSTYRENE
IN PRESENCE OF SULFUR DIOXIDE
Part I. Chain scission as Function of Temperature at Constant
Oxygen and Sulfur Dioxide Pressures and Constant Light Intensity

H. H. G. Jellinek and J. F. Kryman*
Department of Chemistry
Clarkson College of Technology
Potsdam, New York 13676

The reaction of isotactic polystyrene in the presence of sulfur dioxide and oxygen at constant light intensity was studied as a function of temperature. The gas pressures were kept constant, oxygen was present in excess (150mmHg) and the sulfur dioxide pressure was quite small (0.85mmHg). A general survey was presented previously of polymer-gas reactions (air pollutants) such as NO_2 and SO_2 [1]; it was ascertained then, that the presence of SO_2 gives rise or enhances chain scission. The effect of NO_2 on polystyrene and butyl rubber was studied in detail previously[2].

It is shown that sulfur dioxide, even at such low pressures as used here, is essential for effecting or enhancing chain scission in presence of near ultra-violet light. The polymer is not affected at all without this small amount of sulfur dioxide at room temperature (25°C). This fact is of significance for problems of air pollution. The mechanism of the photolysis reaction will be discussed in detail in the second part of this work, where further experimental facts are presented.

Experimental

a) Materials
Isotactic polystyrene was kindly supplied by the Dow
*Present address:Eastman Kodak Comp., Rochester, N.Y.

Chemical Co. (#759-32-10). It contained 20% by weight of atactic isomer, which was removed by extraction with methyethylketone. The isotactic sample was purified by precipitating twice from chloroform solution with hexane at 25°C. It was completely dried in a vacuum oven for 24 hours at 45°C. The intrinsic viscosities were measured in benzene at 30° ± 0. 02°C in Ubbelohde viscometers. All chemicals were of reagent grade quality. SO_2-gas was obtained from Matheson Co. (anhydrous grade, 99.98% SO_2 min., moisture 50 p. p. m., non-volatiles 30 p. p. m., max. acidity as H_2SO_4 10 p. p. m. max.); oxygen was obtained from the same source (extra dry, O_2-99. 6%, N_2 5. 5 p. p. m., H_2O 2. 0 p. p. m., Kr 12. 9 p. p. m., Xe 0 5. p. p. m.)

b) Polymer Film Preparation

Films were cast from 2% w/v solutions of the polymer after filtering chloroform solutions through glass crucibles (porosity 40-60 ASTM). The films were formed on clean mercury surfaces in petri-dishes. The volume used for casting was such that the film thickness was of the order of 25μ to 30μ. The rate of vaporization of the solvent was carefully controlled. The films were placed in a drying chamber attached to a high vacuum apparatus and dried for 48 hours at 60°C (ca. 10^{-5} mmHg) to constant weight.

c) Apparatus

An all glass high vacuum apparatus was used for the investigation. The reaction vessels were arranged in a circle around the U. V. light source. The latter was a high pressure mercury lamp (Hanovia #654-A-36), held by a quartz water cooling jacket; the current was stabilized by a constant voltage transformer (Sola). About five minutes were required for the light intensity to become constant. The reaction vessels were of Pyrex glass (#7740), which eliminated all radiation below 280 mμ. Wavelengths above 360 mμ had transmission values of more than 95%.

The polymer films were mounted in the reaction vessels on glass slides; the latter could be telescoped into a sleeve at the base of the reaction vessel. Thus, the film was located vertically, facing the light source. A linear McLeod gauge was attached to the apparatus. The SO_2 pressure was adjusted to the desired value (ca. 0.8 mmHg) by means of this gauge. The apparatus was heated, before introducing SO_2, by a hot air blower to remove all residual SO_2 adsorbed on the glass walls. Correction for deviations of SO_2 from ideality were too small to be applied[3]. Also the correction for the gas pressure in the apparatus was negligible in this case[4].

Before starting the actual experiments, the films were exposed to SO_2 and O_2 in the dark. It was shown that SO_2 (1 cmHg) has no effect on isotactic polystyrene in absence of light for at least 10 hours at 35°C.

The light intensity was determined using potassium ferrioxalate[5]. Each polymer film received approximately 2.6×10^5 quanta/cm^3 sec. After exposure, the films were degassed for one hour at room temperature in vacuum (ca. 10^{-2} mmHg), to prevent continuation of secondary reactions.

d) Intrinsic Viscosity Measurements and Transformation to Number Average Chain Lengths.

The initial intrinsic viscosity was $[\eta] = 2.23 \frac{dl}{g}$ at $30° \pm 0.02°$ C. The Mark-Houwink equation for benzene solution is[6],

$$[\eta] = 1.06 \times 10^{-4} \overline{M}_v^{0.735} \frac{dl}{g} \tag{1}$$

The sample had a random size distribution of molecular weights[7]. This type of distribution remains invariant for moderate degrees of degradation. The relation between viscosity and number average molecular weights is given for such a case by,

$$\overline{M}_n = \frac{\overline{M}_v}{\left((a+1) \ \overline{|(a+1)}\right)^{1/a}} \tag{2}$$

Here a = 0.735. Thus, the number average chain-lengths can be obtained at any stage of the reaction and hence, also the degrees of degradation $\alpha = \frac{1}{\overline{DP}_{n,t}} - \frac{1}{\overline{DP}_{n,0}}$.

Experimental Results

The polymer films were exposed to SO_2-U.V.-O_2 and constant oxygen (15cmHg) and sulfur dioxide (0.85mmHg) pressures over a range of temperatures from 30°C to 57°C.

The results are given in Fig. 1., where α is plotted versus t; straight lines are obtained. The experimental rate constants, k_{exp}, derived from the slopes are as follows,

°C	30	45	57
$10^6 k_{exp.}$ (hr^{-1})	5.96	7.57	10.50

A good straight line is obtained for the Arrhenius plot; the corresponding equation is,

$$k_{exp} = 1.02 \times 10^{-2} \, e^{-4500/RT} \, hr^{-1}$$

Discussion

It is note-worthy that the vacuum photolysis and photo-oxidation of atactic polystyrene give energies of activation of 2.9Kcal/m and 6.0 Kcal/m, respectively, for $\lambda = 2537A$[8]. SO_2 is essential for the chain scission under the experimental conditions; SO_2 in the dark is ineffective (25°C). SO_2 absorbs near-U.V. light to form an excited singlet, which deactivates within 10^{-8} sec. by fluorescence. In about 1% of the cases a triplet is formed which survives about 10^{-3} seconds. During this time it suffers about 10^6 collisions and can react with SO_2, O_2 or hydrocarbons[9].

A detailed mechanism will be given in Part II of this work. The above reaction scheme leads finally to an expression for the degree of degradation α as follows (see Part II),

$$\alpha = \frac{1}{\overline{DP}_{n,t}} - \frac{1}{\overline{DP}_{n,0}} = K_I(k_1[O_2] + k_2[SO_2]) \, t = k_{exp} t \quad (3)$$

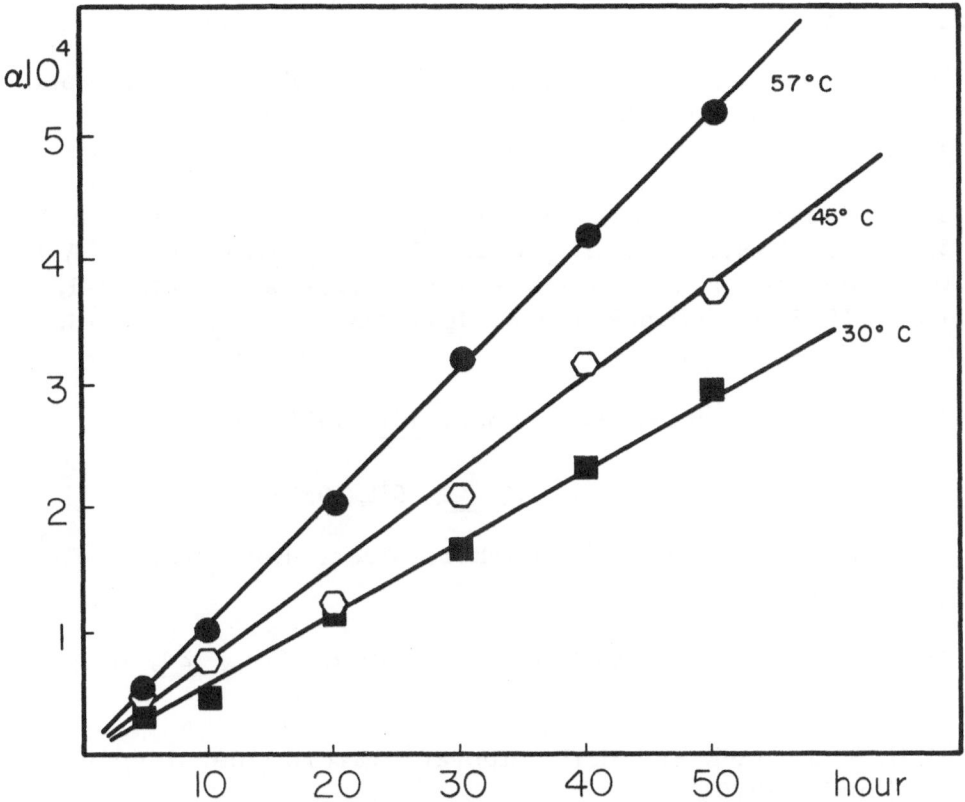

Fig. 1. - Degree of degradation of isotactic polystyrene exposed to U. V. light ($\lambda > 2800A$, $SO_2 \sim 0.85$ mm Hg, $O_2 \sim 15$ cm Hg) as function of time and temperature.

Thus, the degree of degradation is dependent on the oxygen and sulfurdioxide concentrations, respectively; the two rate constants k_1 and k_2, are functions of the light intensity.

This work was supported by a Grant of the Bureau State Services, PHS, Division of Air Pollution (No. 1-RO-1-AP00486).

Summary
    Isotactic polystyrene films exposed to SO_2-U.V.-O_2 ($\lambda > 2800A$) suffer chain scissions according to a random mechanism. The gas pressures ($SO_2 \sim 0.85$ mmHg and $O_2 \sim 150$ mmHg,) were kept constant throughout. This small sulfur dioxide pressure is essential for chain scission to take place. The energy of activation for a combination of secondary reactions is 4.5 Kcal/m.

References:

1) H.H.G. Jellinek, F. Flajsman and F.J. Kryman, J. Appl. Polym. Sci., 13, 107, (1969); H.H.G. Jellinek and F.J. Kryman, ibid., 13, 2504, (1969).

2) H.H.G. Jellinek and Y. Toyoshima, J. Polym. Sci., A-1, 5, 3213, (1967); H.H.G. Jellinek and F. Flajsman, ibid., A-1, 7, 1153, (1969);H.H.G. Jellinek and F. Flajsman, ibid, A-1, in press; H.H.G. Jellinek and S. Igarashi, J. Phys. Chem., April, 1970.

3) M. Francis, Trans. Far. Soc., 131, 1325, (1935).

4) J.A. Barnard, J. Sci. Inst., 34, 511,(1957).

5) G.G. Hatchard and C.A. Parker, Proc. Roy. Soc., A235, 532, (1956).

6) G. Danusso and G. Moraglio, J. Polym. Sci., 24, 161, (1957).

7) H.H.G. Jellinek and S.N. Lipovac, Macromolecules, in press (April, 1970).

8) N. Grassie and N.A. Weir, J. Appl. Polym. Sci., 9, 987, (1965).

9) H.S. Johnston and K. Du Jain, Science, 131, 1532, (1960).

PHOTOLYTIC OXIDATION OF ISOTACTIC POLYSTYRENE
IN PRESENCE OF SULFUR DIOXIDE
Part II. Photolysis Reaction as Function of Light Intensity,
Sulfur Dioxide and Oxygen Pressures

H. H. G. Jellinek and J. Pavlinec*
Department of Chemistry
Clarkson College of Technology
Potsdam, New York 13676

The investigation of this photolysis reaction has been
continued (see Part I[1]), studying the reaction as a function
of gas pressures and light intensity. In the first part of
this work, an energy of activation of 4.5Kcal/m was obtained for constant SO_2 (0.84mmHg) and oxygen (150mmHg)
pressures, respectively. The additional data obtained in
this second part permits the formulation of a mechanism for
this photolysis reaction, which is in accord with all experimental facts; it is based on a type of Bolland mechanism[2]
for oxidative degradation in conjunction with the pronounced
synergistic action of sulfur dioxide.

Experimental

a) Materials, Apparatus and Procedure.
All solvents were of analytical reagent grade; the
isotactic polymer originated from the same sample as used
in Part I. The procedure of making films was the same as
described before[1]. Two polymer samples were prepared
with intrinsic viscosities, $\eta = 2.14$ dlg^{-1} and $[\eta] = 2.09$ dlg^{-1},
respectively, measured in benzene at 30° $\pm$ 0.02°C. Films
were again about 25μ thick. The apparatus and procedure
were the same as described in Part I. The only difference
is that all reactions were carried out at 57°C.   The films

*Present address: Polymer Institute, Slovak Academy of
Science, Bratislava, Czechoslovakia.

were finally dried in high vacuum for 24 hours at 57°C
before exposure. All films were degassed after exposure in
high vacuum to prevent continuation of the reaction in the
dark.

Corrections for non-ideality of SO_2 due to compression in
the McLeod gauge was not applied as it amounts to 0.5%
for the highest SO_2 pressure used. However, a correction
was made for the actual gas pressure in the vacuum
apparatus.[1)]

Number average chain lengths or molecular weights were
derived in the same way as before (Part I), as the samples
had again random molecular size distributions. Straight lines
were evaluated by the "least squares" method.

b) Results

Some preliminary experiments were carried out, repeating
data obtained previously (Part I). The results agree within
about 3% with the corresponding ones of Part I. The following new data was obtained in this work at 57°C:

(1) Degradation does not take place in absence of gases
and U.V. irradiation.

(2) Neither does it occur with U.V. irradiation ($\lambda > 2800A$)
alone in absence of gases.

(3) Chain scission takes place with U.V. irradiation in
presence of sulfur dioxide (3.5mmHg). However, this
degradation stops after an average of 0.2 main chain links
have been severed in each original chain. A film was exposed under these conditions, dissolved and the polymer precipitated. A new film was made from the precipitated polymer
and exposed again under identical conditions as before. Chain
scission did not occur, indicating that inhibitors are not responsible for the cessation of chain scission; rather, a
limited number of "abnormal" (weak) structures must be
present in the original polymer.

(4) Chain scission occurs on exposure to U.V. light in
presence of oxygen alone (150mmHg). Fig. 1 shows the experimental results (1 to 4).

(5) Chain scission is enhanced by the addition of small
amounts of SO_2 at constant oxygen pressure (150mmHg) and
constant light intensity. The degree of degradation

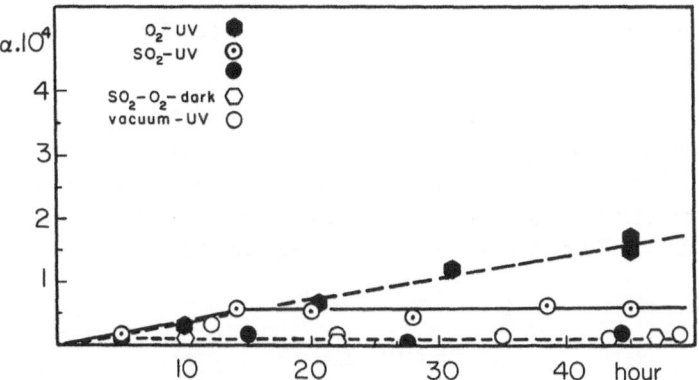

Fig. 1 ⊙ Chain Scission due to "abnormal" (weak) structures in isotactic polystyrene (SO$_2$ 3.5mmHg, U.V. radiation $\lambda \gtrsim 2800$A; 57°C); ● Exposure of newly formed films made from polymer exposed as before (exposed polymer was dissolved, reprecipitated and new films prepared). ◯ Exposure in presence of SO$_2$(3.5mmHg) and O$_2$(150mmHg) and absence of U.V. light (57°C). ◯ U.V. irradiation alone (57°C); ⬡ O$_2$ (150mmHg) and U.V. irradiation (57°C).

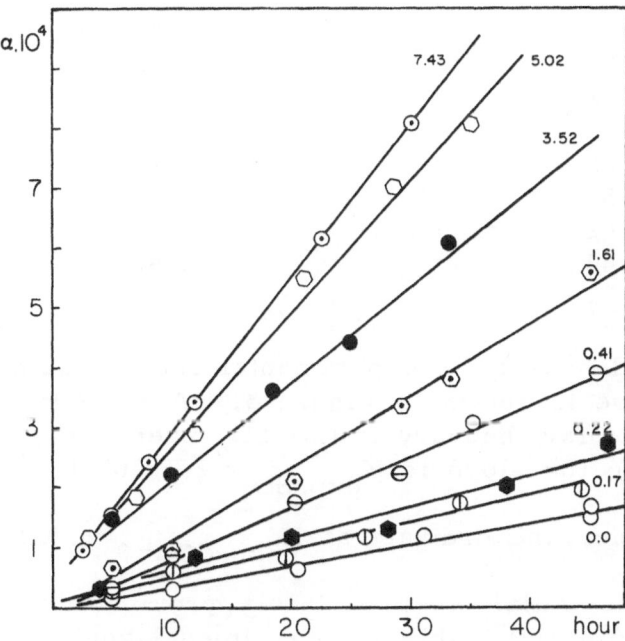

Fig. 2 Degree of degradation, α, as function of SO$_2$-pressure and time at constant O$_2$-pressure (150mmHg) and light intensity (57°C; numbers represent SO$_2$-pressures).

$$\alpha = \frac{1}{\overline{DP}_{n,t}} - \frac{1}{\overline{DP}_{n,0}} \text{ is plotted versus reaction time for}$$

various SO_2 pressures (0 to 7.4mmHg) in Fig 2. Straight
lines are obtained in accordance with the general relationship, $\alpha = kt$, for pure random chain scission. However,
these straight lines do not pass through zero time for higher
SO_2 pressures. The rate constants (slopes) are given in
Table I.

Table I

Chain Scission Rate Constants as a Function of SO_2-Pressure

at Constant O_2-Pressure (150mmHg) and Constant Light

Intensity (57°C).

P_{SO_2} (mmHg)	$10^5 K_{exp}$ (hr^{-1})
0	0.37
0.17	0.43
0.22	0.50
0.41	0.87
0.84*	1.05
1.59	1.23
1.61	1.23
3.50	1.65
3.52	1.62
5.02	2.27
7.43	2.61

* Part I.

The dependence of the experimental rate constants, K_{exp},
on SO_2-pressure is shown in Figure 3. A straight line is
obtained, except for the very initial SO_2-pressure interval
(0 to 0.4mmHg); the slope is $K_{exp, SO_2} = 2.5 \times 10^{-6} hr^{-1} (mmHg)^{-1}$.

The plot intersects the ordinate at $K_{exp, SO_2 = 0} = 8.0 \times 10^{-6} hr^{-1}$.

The slope of the initial steep part is approximately
$1.25 \times 10^{-5} hr^{-1} (mmHg)^{-1}$; this is five times larger than the
value for the straight line at higher SO_2 pressures.

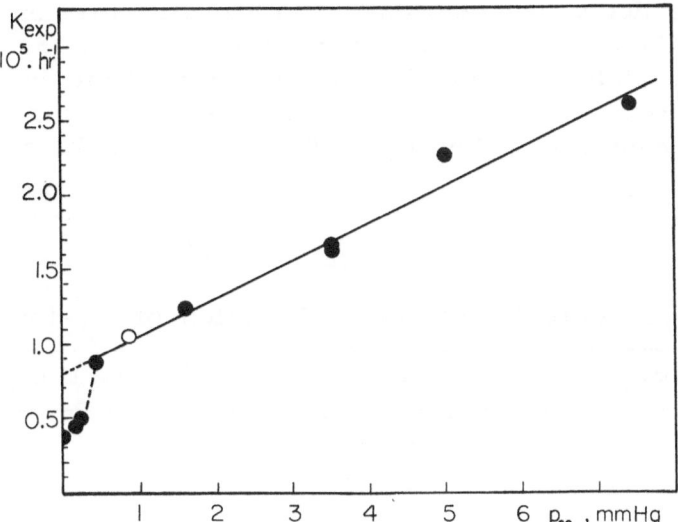

Fig. 3 Chain Scission rate constants, k_{exp}, derived from slopes in Fig. 2 as function of SO_2-pressures at constant O_2-pressure (150mmHg) and light intensity (57°C).

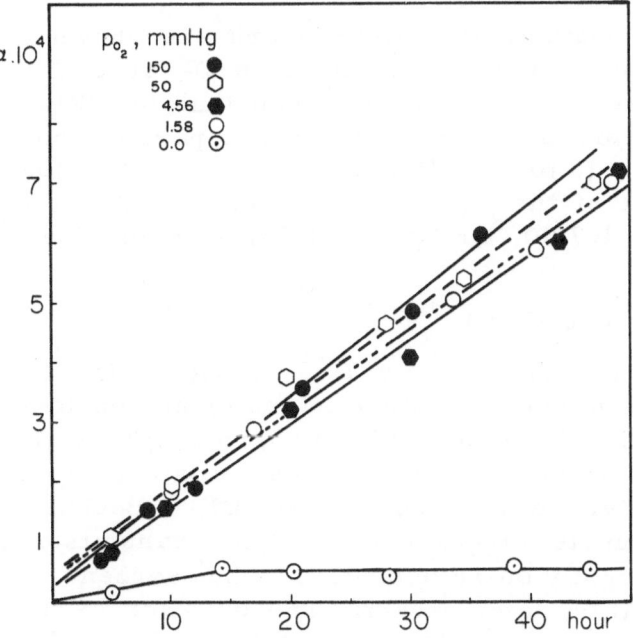

Fig. 4 Degree of degradation, α, as function of O_2-pressure and time at constant SO_2 pressure (3.5mmHg) and light intensity (57°C).

(6) The degree of degradation for chain scission as a function of O_2-pressure at constant SO_2-pressure (3.5mmHg) and constant light intensity is plotted versus time in Figure 4. Here straight lines are also obtained for oxygen pressures from 0 to 150mmHg. Table II gives the relevant rate constants.

<u>Table II</u>

Chain Scission Rate Constants as Function of O_2-Pressure at Constant SO_2-Pressure (3.5mmHg) and Constant Light Intensity (57°C).

P_{O_2} (mmHg)	0	1.58	4.56	50.0	150.0
$10^5 K_{exp}$ (hr^{-1})	0.23	1.36	1.42	1.44	1.65

The rate constants vs. O_2-pressure are shown in Figure 5. The increase in these constants with O_2-pressure at constant SO_2-pressure is much smaller than that for SO_2-pressures at constant O_2 pressure. The very initial part of the plot for small O_2 pressures (0 to 2mmHg) has again a much steeper slope than the rest of the curve. The slope at higher O_2-pressure is $K_{exp, O_2} = 1.7 \times 10^{-8} hr^{-1} (mmHg)^{-1}$ with an intersection on the ordinate of $K_{exp, O_2 = 0} = 1.38 \times 10^{-5} hr^{-1}$. The slope of the initial part is approximately $7 \times 10^{-6} hr^{-1} (mmHg)^{-1}$.

(7) The influence of light intensity at constant gas pressures ($O_2 \sim 150mmHg$; $SO_2 \sim 3.5mmHg$) was also investigated. Metallic wire screens were used to reduce the light intensity. It was shown first (DB-G Beckman Spectrophotometer) that the percentage transmission depends only very slightly on wavelengths. The screens were wrapped around the U.V. lamp jacket. The distance of the lamp from the reaction vessels was about ten times that of the screen from the lamp. The % transmissions of the wire screens were measured and were as follows: 21, 37.5 and 65.2. % transmissions were also calculated from the

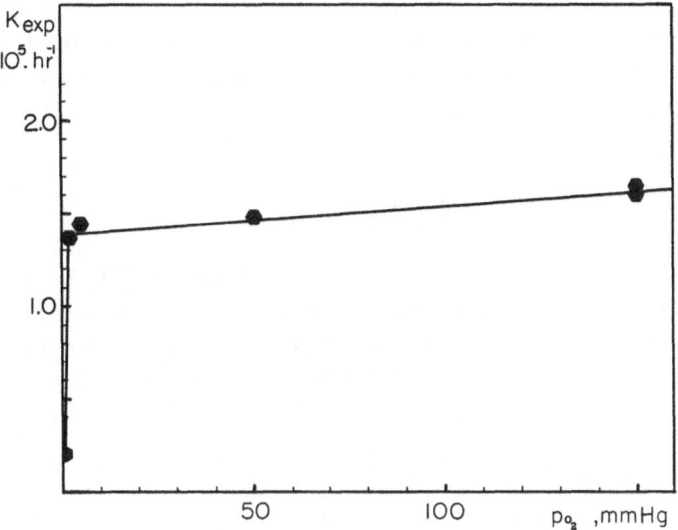

Fig. 5 Chain Scission rate constant, k_{exp}, derived from slope in Fig. 4 as function of O_2-pressures at constant SO_2-pressure (3.5mmHg) and light intensity.

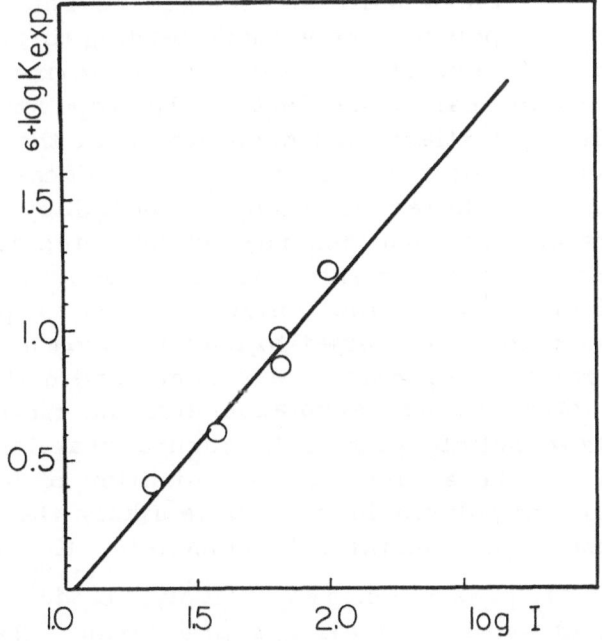

Fig. 6 Logarithmic plot of rate constants versus light intensity in per cent ($O_2 \sim 150$mmHg, $SO_2 \sim 3.5$mmHg, 57°C).

dimensions of the screens; the calculated values agree satisfactorily with the measured data.   Fig. 6 shows the logarithmic plot of the experimental rate constants versus light intensity.   A straight line of slope 1.2 results (least squares).

Discussion

The experimental data enumerated above [(1) to (7)] have to be accounted for in any proposed mechanism.   In addition, it was shown in Part I that for O_2 =150mmHg, SO_2 = 0.84mmHg and U.V. radiation, an Arrhenius equation was obtained for the range of 30°C to 57°C as follows,

$$k = 1.02 \times 10^{-2} \, e^{-\frac{4500}{RT}} \quad hr^{-1}$$

A mechanism for chain scission is proposed here, which accounts satisfactorily for all observed data.   It is based on a type of Bolland[2] mechanism for oxidative degradation, modified by the effect of sulfur dioxide.   Dainton and Ivin[3] showed that sulfinic acid groups in low molecular weight hydrocarbons did not lead to cleavage; accordingly, it was assumed here that these sulfinic acid groups replace tertiary hydrogen atoms in polystyrene without leading to subsequent chain scission.   At low SO_2-pressures, the formation of sulfinic acid groups can be neglected. Hydroperoxide groups, however, replacing tertiary hydrogen atoms in the polymer do not only lead to inert products, but also occasionally to chain scission. The latter reaction is negligible from a chemical point of view, but can readily be detected by molecular weight measurements due to the extended, thread-like nature of the polymer molecules.   The hydroperoxide groups have here to be assumed to lead to inert products due to two parallel reactions, one proceeds with the aid of SO_2 and light, the other occurs spontaneously. The first reaction probably leads to sulfate groups[4]. Such a reaction has been recently used for the analytical determination of hydro-peroxide groups in polyethylene[4]. It is likely that the decrease of the experimental rate constants, K_{exp}, for small SO_2 and O_2 pressures, respectively, is due to diffusion control at these small gas pressures.   The fact that SO_2 alone in presence of U.V. radiation leads to small amounts of chain scission, indicates the presence of a

limited number of especially susceptible groups, probably hydroperoxide groups, in the polymer formed during its history.
The amount of such "weak" links is very much smaller than
in the case of atactic polystyrene[5]; Nakagima et al[6] could
not detect any "weak" links in isotactic polystyrene during
thermal degradation. The observation that a number of
α versus t curves do not pass through the origin of the coordinate system supports the assumption of the presence of
such "weak" links. — It is quite unlikely that any significant
amounts of O_3 are formed under the experimental conditions ($\lambda \not> 2800A$). Taking all experimental facts into
consideration, and taking note that chain scission in highly
viscous media proceeds via medium cages, a mechanism
is proposed as follows:

$$RH + O_2 \xrightarrow{h\nu} R\cdot + \cdot HO_2 \qquad k_1$$

$$RH + SO_2 \xrightarrow{h\nu} R\cdot + \cdot HSO_2 \qquad k_2$$

$$R\cdot + O_2 \longrightarrow \cdot RO_2 \qquad k_3$$

$$\cdot RO_2 + RH \longrightarrow ROOH + R\cdot \qquad k_4$$

$$ROOH \longrightarrow \text{Inert Products} \qquad k_5$$

$$ROOH + SO_2 \longrightarrow \text{Inert Products} \qquad k_6$$
$$\text{(e.g. } ROSO_2OH)$$

$$ROOH + SO_2 \xrightleftharpoons{h\nu} Cage_1 \qquad k_7, k_8$$

$$Cage_1 \longrightarrow \text{Chain Scission} \qquad k_9$$

$$ROOH \rightleftharpoons Cage_2 \qquad k_{10}, k_{11}$$

$$Cage_2 \longrightarrow \text{Chain Scission} \qquad k_{12}$$

$$R\cdot + SO_2 \longrightarrow \cdot RSO_2 \qquad k_{13}$$

$$\cdot RSO_2 + RH \longrightarrow \cdot R + RSO_2H \qquad k_{14}$$

$$RSO_2H \longrightarrow \text{Inert Products} \qquad k_{15}$$

RH and ROOH stand for polymer and polymer hydroperoxide,
respectively; $\cdot R$, $\cdot RO_2$ and $\cdot RSO_2$ are polymer radicals.

The above mechanism leads to an expression for the degree of degradation, α, as a function of reaction time, gas pressures and constant light intensity as follows (see Appendix),

$$\alpha \cong \frac{(k_1[O_2] + k_2[SO_2]) k_3[O_2]}{(k_5 + k_6[SO_2])(k_3[O_2] + k_{13}[SO_2])} \left\{ \frac{k_7 k_9}{k_8} [SO_2] + \frac{k_{10}k_{12}}{k_{11}} \right\} t$$

$$\alpha \cong k_{exp} t \qquad (1)$$

If $[SO_2] < [O_2]$, eq. (1) reduces to,

$$\alpha \cong \frac{k_{10} k_{12}}{k_5 k_{11}} (k_1[O_2] + k_2[SO_2]) t = k_{exp} t \qquad (2)$$

For $[SO_2] = 0$,

$$\alpha \cong \frac{k_{10} k_{12}}{k_5 k_{11}} k_1 [O_2] t = k_{exp} t \qquad (3)$$

For $[O_2] = 0$, $\alpha = 0$ provided no hydroperoxide groups are present at $t = 0$.

If $[SO_2] > [O_2]$,

$$\alpha \cong \frac{k_2 k_5 k_7 k_9 [O_2][SO_2]}{k_8 k_{13}(k_5 + k_6[SO_2])} \, t = k_{exp} t \qquad (4)$$

If there is a definite concentration, $[ROOH]_0$, of hydroperoxide sidegroups in the polymer at $t = 0$, then chain scission proceeds in presence of SO_2 alone and U.V. light. The reaction stops, when all hydroperoxide groups have disappeared. Hence, one has in this case,

$$\alpha = \frac{k_7 k_9}{k_8[n]_0} [ROOH]_0 (1 - e^{-[SO_2]t}) \cong \frac{k_7 k_9}{k_8[n]_0} [ROOH]_0 [SO_2] t$$

$$= k_{exp} t \qquad (5)$$

and $[ROOH]_t = [ROOH]_0 \, e^{-[SO_2] \, t}$ (5a)

This reaction stops when $[ROOH]_t = 0$.

The rate constants k_1, k_2 and k_7 can be expressed as follows,

$$k_1[n]_0 = \phi_{O_2} I_{abs.} \; ; \; k_2[n]_0 = \phi_{SO_2} I_{abs.} \quad \text{and} \quad k_7[ROOH] =$$

$\phi_{ROOH} I_{abs.}$, where ϕ_{O_2} and ϕ_{SO_2} are the chain scission

quantum yields $\left(\dfrac{\text{number of chain scissions}}{\text{number of quanta absorbed}}\right)$ for the reaction

with variable O_2 pressures, constant SO_2 pressure and constant incident light intensity, and variable SO_2 pressures, constant O_2 pressure and constant light intensity, respectively. ϕ_{ROOH} is the quantum yield for $[ROOH]$ moving into a medium cage. $I_{abs.}$ is the light intensity in einsteins absorbed per time unit by each volume unit of a film. $[n]_0 = $ constant is the monomeric unit molar concentration of the polymer.

The ratio, ϕ_{SO_2}/ϕ_{O_2}, of the chain scission quantum yields, can be obtained as follows (see eq. 2),

$$\alpha/t = k_{exp} \cong \frac{1}{[n]_0} \frac{k_{10}k_{12}}{k_5 k_{11}} I_{abs.} \, (\phi_{O_2}[O_2] + \phi_{SO_2}[SO_2])$$

          (6)

For $[O_2] = $ constant and for two pressures of SO_2, one obtains,

$$k_{exp, SO_{2,2}} - k_{exp, SO_{2,1}} = \phi_{SO_2}([SO_2]_2 - [SO_2]_1)\frac{k_{10}k_{12}I_{abs.}}{k_5 k_{11}[n]_0}$$

          (6a)

and for $[SO_2] = $ constant and two O_2-pressures,

$$k_{exp, O_{2,2}} - k_{exp, O_{2,1}} = \phi_{O_2}([O_2]_2 - [O_2]_1)\frac{k_{10}k_{12}I_{abs.}}{k_5 k_{11}[n]_0} \quad (6b)$$

Hence,

$$\phi_{SO_2}/\phi_{O_2} = \frac{(k_{exp, SO_2, 2} - k_{exp, SO_2, 1})([O_2]_2 - [O_2]_1)}{(k_{exp, O_2, 2} - k_{exp, O_2, 1})([SO_2] - [SO_2]_1)} \quad (7)$$

The average value for this ratio calculated from several pairs of k_{exp} values is 143 with a standard deviation of $\pm$ 13. The very large quantum yield for SO_2 (143 times larger than that for O_2) accounts for the large effect of even small pressures of SO_2 on the chain scission reaction. The light intensity exponent is somewhat larger than one (i.e. 1.2). This may be due to the reaction of hydroperoxide groups with SO_2 in the presence of U. V. radiation (reaction 7). Inspection of eq. (1) shows that for this general equation, k_{exp} is dependent on a power of light intensity larger than one. It is possible that $k_7[ROOH] = \phi_{ROOH}I_{abs.}$ has still some influence on eq. (2).

The energy of activation of 4.5Kcal/m belongs to the expression $k_{10}k_{12}/k_{11}k_5$. The small value of the pre-exponential factor (ca. 10^{-2} hr^{-1}) may be in part due to the fact that it contains the quantum yields.

The proposed mechanism accounts satisfactorily for all experimental facts so far known; it shows clearly the pronounced synergistic effect of sulfur idoxide in this chain scission reaction.

Acknowledgement: This work was supported by a Grant of the Bureau of State Services, PHS, Division of Air Pollution (No. R-501-AP-00486-04).
Summary
      The chain scission reaction of isotactic polystyrene has been studied as a function of sulfur dioxide and oxygen pressures, near-U. V. light intensity ($\lambda \neq 2800A$) and temperature. This reaction can be satisfactorily accounted for by a mechanism where chain scission proceeds via the formation of hydroperoxide groups. This chain scission is enhanced by the synergistic action of sulfur dioxide. The ratio of chain scission quantum yields due to sulfur dioxide

and oxygen, respectively, is very large: ϕ_{SO_2} / ϕ_{O_2} = 143.

This shows clearly the great effect sulfur dioxide has, even at small pressures, on this chain scission reaction.

References

1) H.H.G. Jellinek and F.R. Kryman, Symp. at Anaheim, California, Plenum Press, Part I; also MS Thesis, Clarkson(1969).

2) J.L. Bolland, Quart. Rev., <u>3</u>, 1, (1949).

3) F.S. Dainton and K.J. Ivin, Trans. Far. Soc., <u>46</u>, 379, 382 (1950).

4) J. Mitchell, Jr. and L.R. Perkins, Appl. Polym. Symp., <u>9</u>, 167 (1967).

5) H.H.G. Jellinek, Trans. Far. Soc., <u>40</u> (6) 1, (1944) and J. Polym. Sci., <u>3</u>, 850 (1948).

6) A. Nakagima, F. Hamada and T. Shimuzu, Makrom. Chem., <u>90,</u> 229, (1966).

7) A.M. North, Quart. Rev., <u>20,</u> 421, (1966).

Appendix

The rate of chain scission is given, according to the proposed mechanism, by,

$$- \frac{d[n]}{dt} = k_9 [Cage_1] + k_{12} [Cage_2] \qquad (1')$$

The steady state concentrations of the cages are,

$$[Cage_1] = \frac{k_7}{k_8} [ROOH] [SO_2] \qquad (2a')$$

and

$$[Cage_2] = \frac{k_{10}}{k_{11}} [ROOH] \qquad (2b')$$

Further, the steady state concentrations of $R\cdot$ and $\cdot RO_2$, respectively, amount to,

$$[R\bullet] =$$

$$\frac{k_1[n]_0[O_2] + k_2[n]_0[SO_2] + k_4[n]_0[\bullet RO_2] + k_4[n]_0[\bullet RSO_2]}{k_3[O_2] + k_{13}[SO_2]}$$

The last two terms in the nominator are negligible compared with the first two, hence,

$$[R\bullet] = \frac{k_1[n]_0[O_2] + k_2[n]_0[SO_2]}{k_3[O_2] + k_{13}[SO_2]} \tag{3'}$$

and

$$[\bullet RO_2] = \frac{k_3[O_2]}{k_4} \frac{k_1[O_2] + k_2[SO_2]}{k_3[O_2] + k_{13}[SO_2]} \tag{4'}$$

Further,

$$\frac{d[ROOH]}{dt} = k_4[\bullet RO_2][n]_0 - k_5[ROOH] - k_6[ROOH][SO_2] \tag{5'}$$

Introducing (4') into (5') and integrating, gives,

$$[ROOH] = \frac{k_3[O_2](k_1[n]_0[O_2] + k_2[n]_0[SO_2])}{(k_3[O_2] + k_{13}[SO_2])(k_5 + k_6[SO_2])} \qquad x$$

$$x \quad (1 - e^{-(k_5 + k_6[SO_2])t}) \tag{6'}$$

The exponential term in eq. (6') is negligible; insertion of eq. (6') into eq. (1') and integrating yields finally equation (1) of the paper.

If $[SO_2] < [O_2]$, eq. (1) reduces to eq. (2). If $[SO_2] = 0$, all terms containing $[SO_2]$ become zero and eq. (3) remains. If $[O_2] = 0$, $\alpha = 0$; for $[SO_2] > [O_2]$ eq. (5) is obtained.

FLUORESCENCE PROPERTIES OF VISUAL PIGMENTS[1a]

Anthony V. Guzzo, Gary L. Pool[1b], and Calvin B. Leman

Chemistry Department, University of Wyoming

Laramie, Wyoming   82070

The act of seeing, i.e. vision, requires a chromophore capable of absorbing light. In a continuing series of studies Wald, Hubbard, and co-workers[2] have shown that this visual pigment is formed from the combination of a protein with a particular polyene aldehyde. By far the greatest attention has been centered on the night vision pigment, rhodopsin. Its chromophore is derived from 11-cis retinal by reaction of the aldehyde group with a specific amino group on the protein[3] (although there is some controversy over this point[4]). Further, the spectral properties of rhodopsin are best reconciled with the protonated (or acidified) form of this Schiff base.[5]

In discussing the spectroscopic and photochemical properties of rhodopsin, a number of investigators have suggested these are essentially derived from those of the polyene chromophore and that the protein has little effect. Such an extreme view is useful and, since the retinal derived Schiff bases are relatively easy to prepare, provides a good working hypothesis. It is of interest, then, to see which of the spectroscopic or photochemical properties of the visual pigment are derived from those of the polyene and then how they are perturbed by the presence of the protein. For instance, it has been assumed in studies concerning the potential barrier to photoisomerization of the polyene[6,7] that the barrier is essentially that barrier to be found in the isolated Schiff base (the acidified form) and that the protein contributes little.

In the visual cell electron microscope studies have shown that the pigment molecules are arranged in layers, probably only one molecule thick. Such an arrangement is analogous to the arrangement of chlorophyll molecules in the chloroplast and the possibility

105

of energy migration between rhodopsin molecules immediately
suggests itself. We may argue then that such energy transfer may
also be exhibited by the Schiff base at appropriate concentrations.

The present study was carried out then with this in mind. Our
work on the fluorescence characteristics of rhodopsin[8] and of its
intermediates formed upon bleaching[9] naturally led to a study of
the Schiff bases derived from retinal with particular reference to
the possibility of radiationless energy migration.

EXPERIMENTAL DETAILS

All-trans-retinal and retinol (Distillation Products, Inc.)
were stored in a freezer under a nitrogen atmosphere and were
checked for spectral purity before use. The Schiff bases were
synthesized by mixing an excess of the appropriate amine with a
solution of all-trans-retinal in the presence of K_2CO_3 to remove
the water produced and allow the reaction to proceed to completion.
On formation of the Schiff base the retinal absorption (385 mμ) is
replaced by the imine absorption at 365 mμ. Completion of the
reaction was checked by the absorption spectrum or treatment of the
reaction mixture with Carr-Price reagent which will react with any
unreacted retinal to produce a bright blue color. Protonation of
the Schiff bases to effect bathochromic shifts in the absorption
maxima was achieved by additions of trichloroacetic acid to a
solution of the base in methylcyclohexane. Several different
amines were used and all were similar in their fluorescence characteristics. The data presented here refers to the Schiff base
formed from 1-amino-2-propanol and all-trans-retinal - all-trans-
N-retinylidene-1-amino-2-propanol - which will be abbreviated as
NRAP, the protonated form will be referred to as NRAPH[+].

Because of the low fluorescence yields of the compounds
studied, except for retinol, all spectra were obtained at liquid
nitrogen temperature. Methylcyclohexane was most frequently used
as the solvent since it allowed the larger bathochromic shifts to
be observed upon protonation and formed a reasonably good glass at
low temperatures. Methyltetrahydrofuran was used as the solvent in
the polarization studies since it formed an excellent glass. The
fluorescence spectra were obtained using an Aminco-Bowman spectro-
photofluorometer and were corrected for wavelength response of the
detector (the R136 or 1P28 photomultiplier detectors).

In order to measure the polarization of the fluorescence
emission, the quantity $P = \dfrac{I_{VV} - I_{vH}}{I_{VV} + I_{vH}}$ was measured;[11] here the term

I_{VV} represents the intensity of fluorescence excited by vertically polarized light and measured through a vertical polarizer, I_{VH} is the intensity of fluorescence excited by vertically polarized light but measured through a horizontal filter. Polaroid films HN22 were used to obtain the desired polarizations. No correction terms for instrumental response were applied to this equation since we were interested only in comparisons of polarizations between the various systems investigated, i.e. retinol, the Schiff base (NRAP), and the protonated Schiff base ($NRAPH^+$).

The polarization measurements were obtained at 77° K from cylindrical films formed between the surfaces of a solid glass rod and a sealed off glass tube of slightly larger inside diameter than the solid rod. All polarization values were reproducible to within 10%.

RESULTS AND DISCUSSION

I.   Absorption and Fluorescence Spectra

The fluorescence and excitation maxima obtained for retinol (Table I) correspond with the fluorescence maximum observed by Kahan[12] and the absorption maximum at liquid nitrogen temperature[13] respectively. For retinal the observed fluorescence maximum corresponds to that observed previously[14], allowing for detector response corrections.

All-trans-N-retinylidene-1-amino-2-propanol (NRAP) has an absorption maximum at 365 mμ at room temperature and a fluorescence excitation maximum at ∿375 mμ at -196°C. The bathochromic shift is presumably a result of the temperature difference since it is of the order of magnitude expected[13]. The fluorescence maximum was at 500 mμ at -196°C. The addition of an equimolar quantity of trichloroacetic acid to the Schiff base produces a bathochromic shift in the absorption maximum to 450 mμ (25°C). The excitation and fluorescence maxima at -196°C were 460 mμ and 550 mμ respectively. Additions of a one hundred fold excess of trichloroacetic acid produces a further bathochromic shift in the absorption maximum to 470 mμ with corresponding excitation and fluorescence maxima at -196°C of 480 mμ and ∿590 mμ respectively. The excitation and fluorescence maxima for this series of compounds are also tabulated in Table I.

A. V. GUZZO, G. L. POOL, AND C. B. LEMAN

TABLE I

FLUORESCENCE AND EXCITATION MAXIMA OF RETINOL,
RETINAL, AND MODEL SCHIFF BASES[a]

Compound	Excitation Maximum (mμ)	Fluorescence Maximum (mμ)	$\tilde{\nu}$[b] $\Delta\nu$ (cm^{-1})
Retinol	335	465	8300
Retinal	(385)[c]	520	6800
NRAP	375	500	6700
NRAPH$^+$ (equimolar TCA)[d]	460	550	3500
NRAPH$^+$ (excess TCA)	480	590	3900

[a]All-trans isomers at -196°C in methylcyclohexane solvent.
[b]The difference in the excitation and fluorescence maxima expressed in cm^{-1}.
[c]Because of the distorted excitation spectrum the absorption maximum at -196°C was used.
[d]TCA = trichloroacetic acid.

Table I shows that the energy separation, i.e. the Stokes shift, between the absorption and fluorescence maxima of the unprotonated Schiff base is large and comparable to the same separation in retinal and retinol, whereas the protonated Schiff bases have a considerably decreased separation. This correlates well with the observations on rhodopsin and its bleaching intermediates[8,9] given in Table II. In this series rhodopsin, lumirhodopsin, and metarhodopsin I have small separations whereas metarhodopsin II reverts to a large separation comparable to retinal, retinol, and the unprotonated Schiff base. According to present understanding of the bleaching cycle[2], metarhodopsin II is the first species in the cycle which contains the deprotonated Schiff base linkage. Thus the decrease in the absorption-fluorescence separation in both the rhodopsin system and in the model Schiff bases apparently arises from the protonation.

TABLE II

FLUORESCENCE AND ABSORPTION MAXIMA OF BOVINE RHODOPSIN AND
SOME OF THE INTERMEDIATES FORMED ON BLEACHING[a]

Species	Absorption Maximum (mμ)	Fluorescence Maximum[b] (mμ)	$\widetilde{\nu}$[c] $\Delta\nu$ (cm^{-1})
Rhodopsin	500	600	3300
Lumirhodopsin	497	600	3300
Metarhodopsin I	478	580	3500
Metarhodopsin II	380	535	7600

[a]References 8, 9.
[b]Observed at -196°C from sample of rod outer segments.
[c]The separation between the absorption maximum and the
fluorescence maximum expressed in wavenumbers.

TABLE III

POLARIZATION VALUES OBTAINED FROM THIN FILMS OF RETINOL,
NRAP, AND NRAPH$^+$[a]

Concentration	Retinol	NRAPH$^+$	NRAP
less than 10^{-5}M	60		
2.7×10^{-1}M		5	20
2.7×10^{-2}M		23.6	34.8
2.7×10^{-3}M		40	43.5
2.7×10^{-4}M		37.4	41.5

[a]All values are in percentages and were obtained from the
expression $P = 100 \times \dfrac{I_{vv} - I_{vH}}{I_{vv} + I_{vH}}$.

II. Polarization Studies

There are several ways of estimating the extent of energy transfer between a possible donor and acceptor[15] which, in the systems studied here, are the same molecule. In our system we are limited by the intense absorption of the polyene chromophore and thus decreases in fluorescence intensity cannot be reliably used as a measure of transfer. Further, lifetime studies are experimentally quite difficult. Polarizations techniques appeared to provide the best way of estimating transfer.

If a fluorescence is excited with polarized light we expect a retention of polarization in the emitted light if 1) no molecular tumbling has occurred in the lifetime of the excited state, 2) scattering is negligible, and 3) energy transfer has not occurred.[16] Since our work was done at 77°K, the first point is of little consequence. Scattering is impossible to eliminate but can be accounted for by using some standard that does not exhibit energy transfer. In our case retinol served as the standard since we obtained no evidence of transfer with it except at very high concentrations. The third point--the existence of energy transfer-- could then be determined from the retention of polarization measured in our systems at various concentrations.

In discussing transfer the "critical" transfer concentration C_o and "critical" transfer distance R_o are quantities often used. These terms refer to that concentration and that average interchromophore distance at which energy transfer is equally as probable as all other means of deactivation of the excited state. The two quantities are related by the expression[15]

$$C_o = \frac{397}{R_o{}^3} \dots \dots \dots \dots (1)$$

Thus assuming no molecular tumbling, and zero scattering we expect a fifty percent retention of polarization if energy migration occurs in a system at the critical concentration.

The polarization results obtained from thin films of retinol, NRAP, and NRAPH[+] in methyltetrahydrofuran are given in Table III. The result for retinol was obtained for the lowest concentration of this compound that still allowed a reasonable spectrum to be recorded. As discussed in the previous section, the energy separation between the absorption and fluorescence maxima of retinol is quite large and energy transfer is not expected to occur. Thus a polarization value of 60% indicates instrumental scattering of at least 40%. For the most dilute samples of the Schiff base and its protonated form, the maximum polarization values obtained are near 45%. This value would suggest instru-

mental scattering of at least 55%. There is thus uncertainty in
this point; nevertheless, in the following we will use both of
these values and hopefully obtain a range of "critical" transfer
distances.

    If energy transfer occurs, then at the critical concentration
the polarization should fall to one half its value in the absence
of transfer. For the scattering limit determined by the retinol
case this is P = 30%; while for the scattering limit determined
from the Schiff base study this is near 23%.

    The polarization values obtained for $NRAPH^+$, which are plotted
in Figure 1, give for P = 30% and then 23% "critical" concentrations
of 0.012M and 0.027M respectively. Using equation 1 these "critical"
concentrations give R_o values between 25Å and 32Å, values typical
for singlet-singlet transfer[15] (a similar analysis for the unprotonated Schiff base gives values near 15Å for R_o indicating very
inefficient transfer as expected).

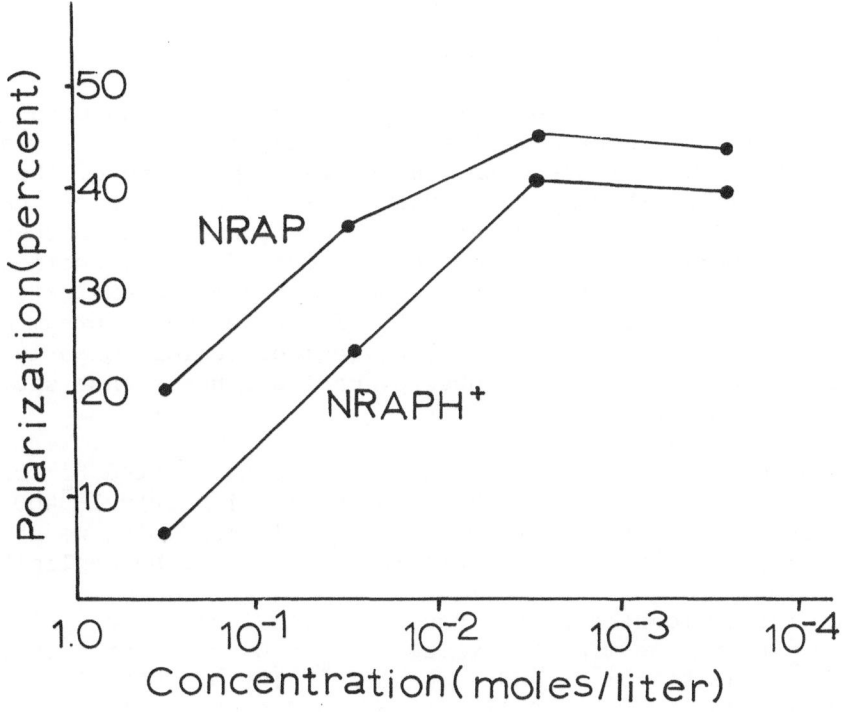

Figure 1. Polarizations of Thin Films of NRAP and $NRAPH^+$ Versus
Concentration.

We may now return to rhodopsin and, considering its protonated Schiff base nature, suggest that R_o values in the 25-32Å range should also hold for its transfer characteristics. The absence of photodichroism in rod outer segments has been observed by Hagins and Jennings[17] and more recently by Pak and Helmrich.[18] Results in this laboratory using the fluorescence depolarization technique strongly suggest that this is indeed an energy migration phenomenon.[19]

III. Spectral Overlap Predictions of Transfer

In our study of the depolarization of the rhodopsin fluorescence[19] we were unable to obtain accurate polarization values and could not then calculate R_o. We may calculate this quantity following Förster's results for dipole-dipole or resonant transfer[20]. Förster's equation in the form used by Tweet, Bellamy, and Gaines[21] is

$$R_o^6 = \frac{9000 \ln 10 <k^2> \eta_o}{128\pi^6 \, n^4 \, N\tilde{(\nu)}_{av.}^{+4}} \int_0^\infty e(\tilde{\nu}) \, f(\tilde{\nu}) \, d\tilde{\nu} \ \ldots\ldots (2)$$

where η_o is the quantum yield of fluorescence in the absence of transfer, n is the index of refraction of the medium, N is Avagadro's number, $<k^2>$ is an average over the angular dependence of the dipolar interaction, $\langle\tilde{\nu}\rangle$ is an appropriate average wavenumber usually found as the average position of the absorption and fluorescence maxima; and the integral is over the product of the absorption spectrum of the acceptor (given as the molar decadic extinction coefficient versus wavenumber) and the fluorescence spectrum of the donor normalized to unity on a wavenumber scale. Clearly the more widely separated are the absorption and fluorescence spectra the less probable will be transfer.

Using data obtained previously[8] we have calculated R_o for rhodopsin. With $\eta_o = 0.005$, $<k^2> = 0.66$, $n = 1.2$, $\langle\tilde{\nu}\rangle = 18,400$ cm^{-1}, and a calculated overlap integral of 4,300 liters/mole, we obtain $R_o = 11$ Å; a value which would predict transfer to be negligible.

For the protonated Schiff base we may calculate an R_o in the same way. Using the measured value of $\eta_o = 0.008$, $<k^2> = .66$, $n = 1.2$, $\langle\tilde{\nu}\rangle = 19,400$ cm^{-1}, and a measured value of 8600 liters/mole for the overlap integral, R_o is again found to be 11 Å. Again we obtain a value of the transfer distance which is too small for transfer to be important.

Of the factors that influence these results, the quantum efficiency of fluorescence η_o is the only one that can vary enough

to make a large difference in calculated result. It is clear that
the very low quantum yield measured for the NRAPH[+] and rhodopsin
fluorescences have produced these low "critical" transfer distances.

We are then faced with a contradiction. It is difficult to
accept that our measured depolarization for both the protonated
Schiff base and rhodopsin are greatly in error, especially in the
face of the lack of transfer we do seem to observe for the unprotonated Schiff base and for metarhodopsin II[19]. Förster's
equation has been verified for a number of systems and appears
valid. However, his analysis was developed for excited states that[22]
had reached thermal equilibrium much before the process of emission.
There is some evidence that this may not be the case, at least for
retinal. Becker[14] has shown that the fluorescence yield of retinal
is wavelength dependent thus indicating the presence of processes
competing with thermal relaxation in the excited state. If similar
phenomena existed within the protonated Schiff bases and in rhodopsin
we would expect that Förster's resonant transfer approach would not
apply.

SUMMARY

Our experiments indicate that the fluorescence changes
observed as rhodopsin bleaches to its various intermediates are
entirely consistent with a change from a protonated to an unprotonated polyene Schiff base. In the protonated form both rhodopsin
and the Schiff base have extensively overlapping absorption and
fluorescence spectra indicating the possibility of radiationless
migration of energy between chromophores. Polarization studies on
rhodopsin and the protonated Schiff base do give evidence of such
transfer. However, if we attempt to calculate a "critical" transfer distance using the theoretical expression obtained by Förster,
we obtain values which are much lower than the measured quantity.
It is our belief that Förster's equation does not apply to these
polyene systems since transfer from the excited vibronic levels
may take place. Since Förster's theory was developed assuming a
thermally equilibrated excited state where only the lowest level
is populated it may not be possible to calculate transfer distances
in this manner. If this is the case then these higher vibronic
levels may be easily populated and transfer rates from them may be
considerably different from those at the lowest excited level.

BIBLIOGRAPHY

1. a) We gratefully acknowledge support for this work by the
    Atomic Energy Commission through contract No. AT(11-1) - 1627;
    b) NIH Predoctoral Fellow (1969) supported through grant No.
    1 - Fl - GM - 36, 155-01: Now at Chemistry Department,
    University of Minnesota.

2. G. Wald, P. K. Brown, and I. R. Gibbon, J. Opt. Soc. Amer.,
    53, 20 (1963); also R. Hubbard, J. Biol. Chem., 241, 1814
    (1966).

3. R. A. Morton and G. A. J. Pitt, Biochem. J., 59, 128 (1955).

4. E. W. Abrahamson, R. P. Poincelot, P. G. Millar, and R. L.
    Kimbel, Jr., Nature, 221, 256 (1969).

5. H. J. A. Dartnall, The Visual Pigments, 1957, John Wiley,
    New York.

6. K. Inuzuka and R. S. Becker, Nature, 219, 383 (1968).

7. J. R. Wiesenfeld and E. W. Abrahamson, Photochem. Photobiol.,
    8, 487 (1968).

8. A. V. Guzzo and G. L. Pool, Science, 159, 312 (1968).

9. A. V. Guzzo and G. L. Pool, Photochem. Photobiol., 9, 565

10. J. O. Erickson and P. E. Blatz, Vision Res., 8, 1367 (1968).

11. A. C. Albrecht, J. Mol. Spect., 6, 84 (1961).

12. J. Kahan, Acta Chem. Scand., 21, 2515 (1967).

13. L. Jurkowitz, J. Loeb, P. Brown, and G. Wald, Nature, 184,
    614 (1959).

14. D. E. Bolke and R. S. Becker, J. Amer. Chem. Soc., 90, 6710
    (1968).

15. N. Turro in Molecular Photochemistry, W. A. Benjamin, Inc.,
    New York, 1967, Ch. 5.

16. E. Gaviola and P. Pringsheim, Z. Physik, 24, 24 (1924).

17. W. A. Hagins and W. H. Jennings, Disc. Fara. Soc., 27, 180
    (1959).

18. W. L. Pak and H. G. Helmrich, Vision Res., 8, 585 (1968).

19. C. B. Leman, Ph.D. Thesis, University of Wyoming, 1969.

20. Th. Förster, Fluoreszenz Organisher Virbindungen, Vandehoeck
     and Ruprecht, Gottingen, 1951.

21. A. G. Tweet, W. D. Bellamy, and G. L. Gaines, Jr., J. Chem.
     Phy., 41, 2968 (1964).

22. Th. Förster, Disc. Fara. Soc., 27, 7 (1959).

SURFACE-PHOTOPOLYMERIZATION OF MALEIMIDES

(Mrs.) A. Christopher, A.K. Fritzsche, and A. Nelson Wright

General Electric Research & Development Center

Schenectady, New York 12301

INTRODUCTION

It has been known for some time that butadiene in contact with metallic substrates polymerizes under the influence of ultraviolet light[1]. Depositions on various substrates from C_4H_6[2,3], and other diene and vinyl monomers[3], and from the perhalogenated monomers C_4Cl_6 and C_2F_4[2-4] have been described in some detail. The surface-photopolymerization process has also been extended to the vapors from solid monomers such as pyromellitic diimide and succinimide[5]. This paper describes surface-deposition of polymeric films from solid maleimide and other imide monomers, with emphasis on the thermally stable films deposited from the vapor of N-phenylmaleimide.

A number of workers have shown that substituted maleic anhydrides and maleimides dimerize readily under UV irradiation to give cyclo-butyl derivatives[6]. Photopolymerization of N-methylcitraconimide[7] and dye-sensitized solid-phase photopolymerization of acrylamide[8] has been reported, as has radiation-induced polymerization of maleimide[9] and of N-p-nitrophenylmaleimide or N-carbamylmaleimide[10]. Yamada, Takase, and Koutou have described solid-state and solution photopolymerization of maleimide and its N-substituted derivatives[11]. Some polymer appears to be produced from (2300-2600 Å) photolysis of succinimide vapor[12].

EXPERIMENTAL

The solid re-crystallized monomers were provided by Dr. F.F. Holub of this Laboratory. Depositions were made as

TABLE I: DEPOSITION FROM VARIOUS IMIDES

Compound	Structure	Mol. Wt.	Cap. (nF)	Thickness (Å)	Growth Rate (Å/min.)
N-phenylmaleimide		173	0.35	≤10,000	≤1000[a]
p-tolylmaleimide		187	0.50	6,720	672[a]
N-phenylphthalimide		223	7.10	470	47[a]
N-phenyltetrahydro-phthalimide		227	1.02	3,300	220[b]
N-vinylphthalimide		173	0.40	8,400	560[b]
N-allylphthalimide		187	0.55	6,345	423[b]
Phenyl imide of 5-norbornene-2,3 dicarboxylic anhydride		241	1.80	1,875	125[b]

a: Deposition Time = 10 min.

b: Deposition Time = 15 min.

described previously, with the quartz-windowed reaction chamber pre-evacuated to a pressure of about 10^{-6} torr before irradiation[2-4]. Under the non-controlled temperature conditions employed in these experiments the temperature of the substrate surfaces increased to about $175^{\circ}-210^{\circ}C$ during deposition. A 700 watt Hanovia medium pressure mercury arc was used as the light source. The monomer was shielded from the light and heated by a heat gun to provide a vapor pressure of about two hundred microns. Substrates for the thin polymeric films included 1 by 3 inch aluminum pieces and aluminum or gold films evaporated onto glass microscope slides.

Measurements of the capacitance and dissipation factor of the thin films were made at 1Kc/s by the mercury drop technique, or by using the film as the dielectric material in a cross-strip capacitor[3].

Infrared spectra of the thin polymeric films (on evaporated Al substrates) were measured by the multiple reflection technique. Infrared and other analytical studies were also made on polymer removed from the inside surface of the quartz reaction vessel by treatment with dilute hydrofluoric acid. Unpolymerized monomer condensed on the inside of the quartz tube could be removed first by dissolving in acetone.

RESULTS - DISCUSSION

Table I lists the imides investigated, their structures, molecular weights and the characteristics of the polymeric films produced from them. Growth rates were obtained from irradiation times and film thicknesses as given by capacitance measurements. The dielectric constant of the polymeric film was determined by combined capacitance and interferrometric measurements of thickness. The value of $\in$ obtained for the polymeric film from N-phenylmaleimide, 3.8 (with a refractive index equal to 1.62), was also used in calculating thicknesses for the other imide polymers.

It is likely that higher growth rates would be obtained for film depositions shorter than 10-15 minutes, since growth of polymer did occur on the inside surface of the quartz reaction vessel. Since both the monomer and polymer absorb in the UV region, Figure 1, the amount of UV radiation reaching the substrate surface is continually decreasing, and hence the growth rate of polymer on the substrate also decreases with increasing times of irradiation. It may be noted that the monomer of lowest growth rate, N-phenylphthalimide, is the only imide investigated in which there is no carbon-carbon double bond external to an aromatic ring.

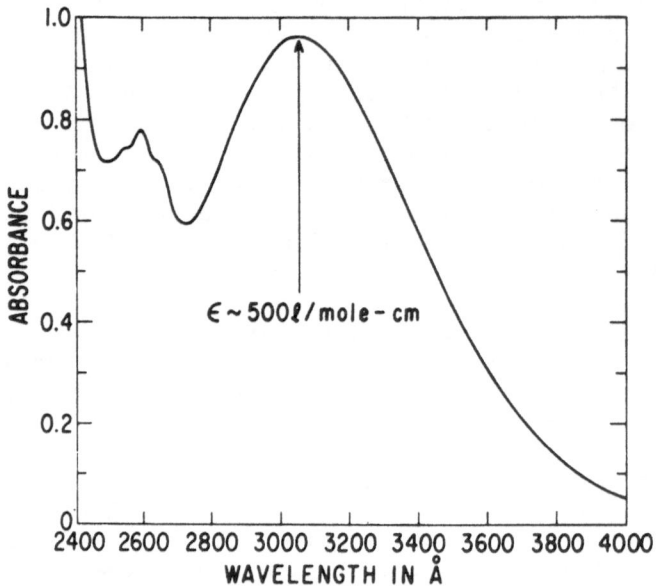

Fig. 1(a) UV absorption spectrum of N-phenylmaleimide,
3.78 mg./10 ml. acetonitrile, 1 cm. path length.

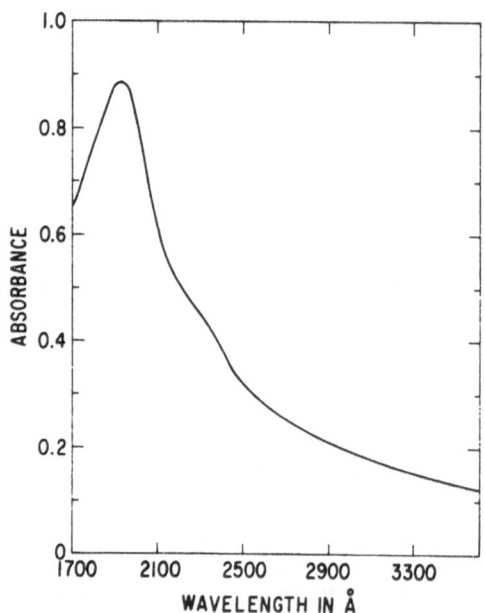

Fig. 1(b) UV absorption spectrum of polymeric film from
N-phenylmaleimide, about 500 Å thick on GE 151 quartz.

If the photopolymers are formed by chain-growth at an olefinic bond, the monomer N-phenylmaleimide would be expected to give the polymer:

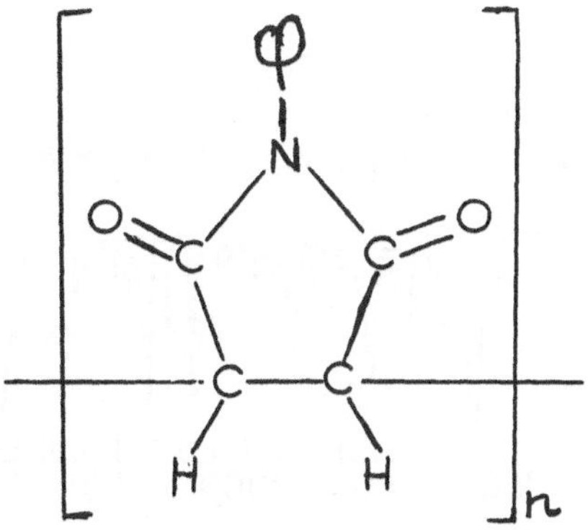

The IR spectrum of the monomer shows a medium-strong band at 830 cm^{-1}, assigned to a carbon-hydrogen (-CH=CH-) deformation mode which is indeed missing from the IR spectrum (multiple reflection method) of the polymer, Figure 2. Otherwise, the spectrum of the polymer film is remarkably similar to that of the monomer. This indicates that the major change in chemical structure during formation of the polymer is the opening of the carbon-carbon double bond. The elemental analysis calculated for the polymer on the basis of this scheme of polymerization is shown below together with the measured values.

ELEMENT	CALCULATED	MEASURED
C	69.4%	68.1%
H	4.1%	3.4%
N	8.1%	7.8%

Deviation from calculated values might be due to side reactions.

Mass spectrometric identification of acetylene, carbon monoxide and carbon dioxide as the major gas phase products, along with smaller amounts of ethylene, nitrogen and methane, does suggest that the photopolymerization may be carried in part by species such as $\phi N(CO)_2$ formed by elimination of C_2H_2. Ethylene could then result from hydrogenation of C_2H_2. The polymer might then correspond to a certain extent to a random copolymer of the type:

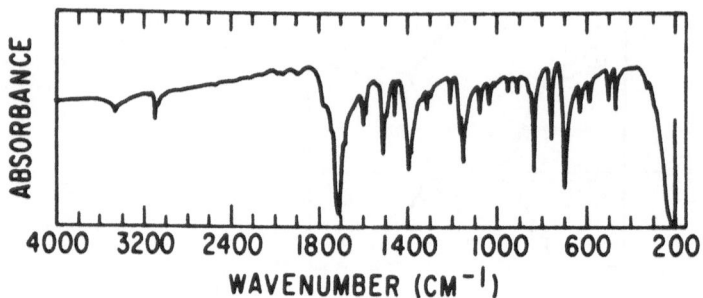

Fig. 2(a)   IR absorption spectrum of N-phenylmaleimide,
            4.82 mg/gm. KBr.

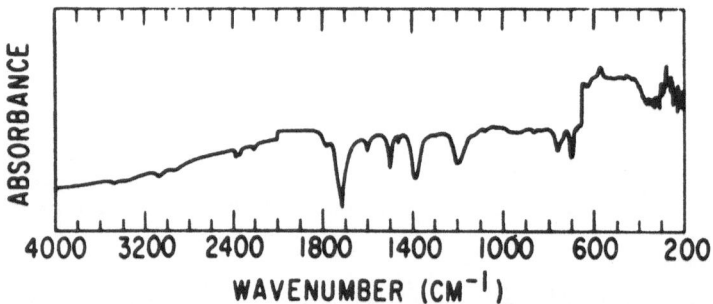

Fig. 2(b)   IR absorption spectrum of polymeric film
            from N-phenylmaleimide, about 2500 Å thick,
            11-13 reflections.

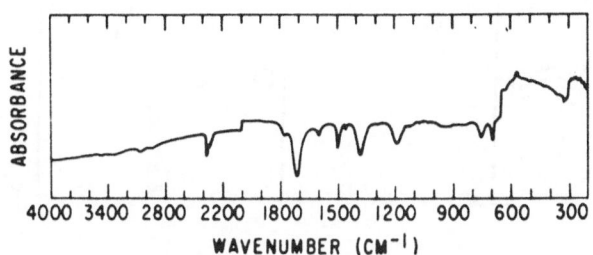

Fig. 3 IR absorption spectrum of polymeric film from
N-phenylmaleimide, about 2500 Å thick, after heating
in air at 300°C for 15 hours, 11-13 reflections.

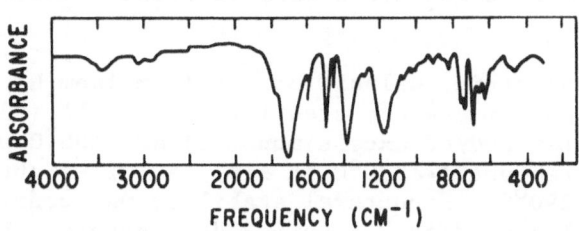

Fig. 4(a) IR absorption spectrum of polymeric film from
N-phenylmaleimide, from quartz surface, 6.33 mg/gm. KBr.

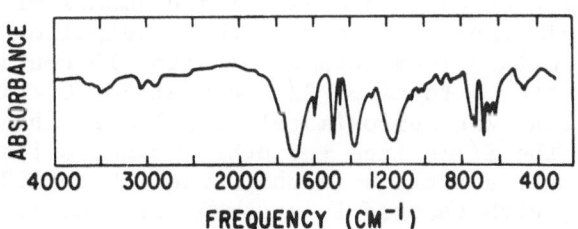

Fig. 4(b) IR absorption spectrum of polymeric film from
N-phenylmaleimide from quartz surface after heating in
air at 300°C for 15 hours, 1.65 mg/gm. KBr.

Such a structure could decrease the tendency for the crystalline monomer to form crystalline polymer and hence account for the absence of first-order transitions on differential scanning calorimetric analysis over the temperature range -100°C to 400°C. Two "second-order" transitions were detected, however, at ~45°C and ~138°C.

Thermogravimetric analysis of the film from N-phenylmaleimide indicated a high temperature stability: a small (~2%) weight loss (presumably water and/or excess monomer) at ~100°C was all that could be detected until ~370°C in air. A 10% weight loss was not induced until 390°C. Structural stability was confirmed by in-frared absorption studies. Figure 3 demonstrates that no obser-vable change* occurs in the spectrum of the film after heating to 300°C in air for 15 hours. The IR spectrum of polymer removed from the inside of the quartz reaction tube, before and after heating to 300°C is shown in Figure 4.

Dielectric studies further revealed the thermal stability of the films from N-phenylmaleimide. Measurements with a mercury drop counter electrode showed that the dissipation factor at room temperature remained less than 0.20% after 15 hours heating in air at 200°C and after a total of 22 hours at 300°C. The apparent di-electric constant was approximately doubled by this heat treatment. In contrast, the films from the other monomers listed in Table I reached dissipation factors in the range 1.2 to 17% after treat-ment at 200°C, with changes in apparent dielectric constant $\geq$ 6.

---

*The apparent peak around 2350 cm^{-1} is a background artifact of the multiple reflection technique.

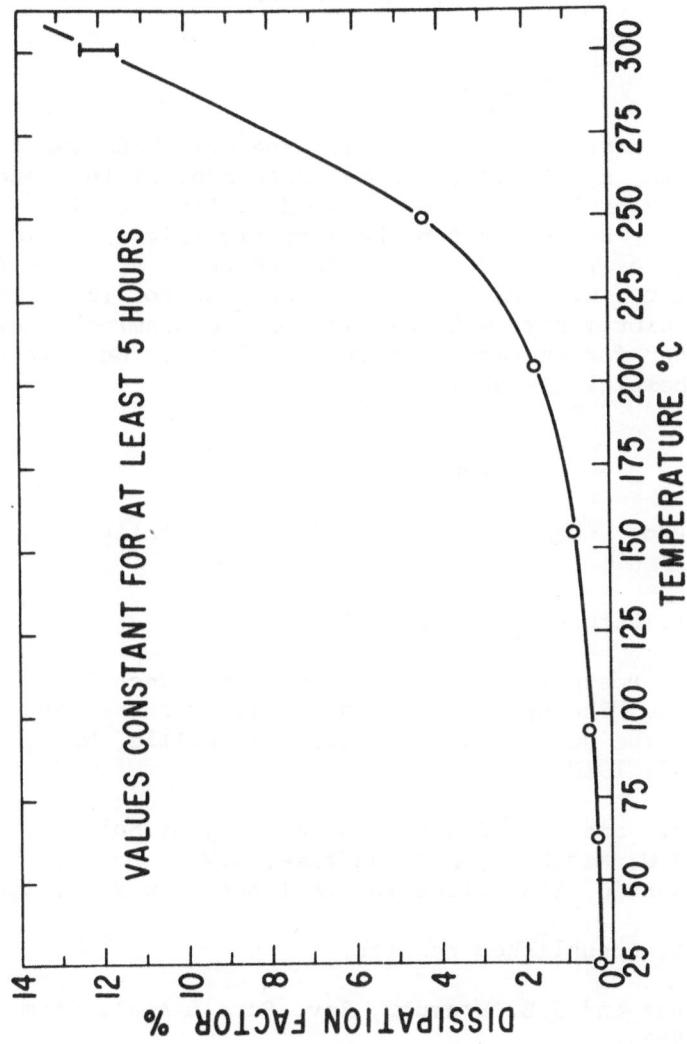

Fig. 5   Temperature dependence of dissipation factor of films
from N-phenylmaleimide.

Figure 5 illustrates "equilibrium" values for the dissipation factor, constant for at least 5 hours, measured at temperatures to 300°C with crossed-film counterelectrodes evaporated over the films from N-phenylmaleimide. Although these measurements are complicated, with values maxima due to experimental difficulties with connections, solder, etc. on exposure to these elevated temperatures, the dissipation factor consistently returned to values ≤ 0.40% in cooling to room temperature.

ACKNOWLEDGMENTS

We are very grateful to F.F. Holub and R.W. Smearing for supplying us with, and in fact synthesizing many of the monomeric imides and for helpful discussions and suggestions during this work. We thank L. Bronk for the thermogravimetric analysis, R.M. Chrenko and D. Temple for the infrared work, A. Holik for the interferromety, N.A. Parker and P. Gundlach for the differential scanning calorimetry, W.G. Racela for the chemical analysis, and G.P. Schacher for the mass spectral analysis. We also thank C.O. Kunz for helpful discussions.

REFERENCES

1. P. White, Proc. Chem. Soc., 337 (1961); Insulation, 13 (5), 52 (1967).

2. A.N. Wright, Nature, 215, 953 (1967).

3. A.N. Wright, Postprints of the SPE Regional Technical Conference on "Photopolymers-Principles, Processes and Materials", The Nevile Country Club, Ellenville, N.Y., November 6-7, 1967.

4. V.J. Mimeault and A.N. Wright, "Reactivity of Solids", Edited by J.W. Mitchell, R.C. DeVries, R.W. Roberts, and P. Cannon, p. 543, Wiley-Interscience, New York, 1969.

5. A.N. Wright, unpublished results.

6. R.N. Warrener and J.B. Bremner, Rev. Pure & Appl. Chem., 16, 117 (1966).

7. T.V. Scheremeteva, M.G. Zhenevshaya, Yu. D. Kondrashev, and Yu. G. Baklagina, Izvest. Akad. Nauk SSSR, Ser. Khim, 1723 (1966).

8. I.M. Barkalov, V.A. Benderskii, V.I. Gol'danskii and
    S.S. Kuz'mina, Doklady Akad. Nauk SSSR, Ser. Khim,
    169, 1111 (1966).

9. T. Kagiya, M. Izu, S. Kawai, and K. Fukui, J. Poly. Sci.,
    B4, 387 (1966).

10. V.S. Ivanov, V.K. Smirnova, and I.R. Kalandadze, Vestn.
    Leningr. Univ. Ser. Fiz Khim, 23, 142 (1968); Vysokomol
    Soedin Ser. B, 151 (1967).

11. M. Yamada, I. Takase, and N. Koutou, Polymer Letters, 6,
    883 (1963).

12. G. Choudhary, A.M. Cameron, and R.A. Back, J. Phys. Chem.,
    72, 2289 (1968).

THE EFFECT OF SOLUTION COMPONENTS ON POLYACRYLAMIDE GEL FORMATION VIA RIBOFLAVIN-SENSITIZED PHOTOPOLYMERIZATION

Howard L. Needles

Department of Consumer Sciences, University of

California, Davis, California   95616

Since Davis (1) first described the use of riboflavin-sensitized photopolymerization of acrylamide-N,N'-methylenebis-acrylamide (BIS) mixtures to form large-pore stacking gels for electrophoresis discs, several authors (2-5) have used this photo-polymerization technique to form gels which are useful for various specific separations and applications. Although extensive use of photopolymerized gels in separations has been made, little is known concerning the effect of commonly added components; i.e., accelerators, buffers, viscosity improvers, leading and trailing ions, and proteins, on the rate and course of photopolymerization. Furthermore, Oster (6, 7) has shown that both oxygen and hydrogen donors effect riboflavin-sensitized photopolymerizations of aqueous acrylamide, and we recently reported (8) that both proteins and amino acids have a dramatic effect on riboflavin-sensitized photo-polymerizations. Since workers are often puzzled by their inability to reproducibly and consistently photopolymerize low concentrations of acrylamide-BIS solutions to gels, we have determined the effect of typical solution additives on formation of polyacrylamide gels. Additives, alone and in combination, are shown to have a marked effect on the photopolymerization rate and gelation time of polyacrylamide-BIS gels.

METHODS

The sources of materials (1, 8) and photopolymerization techniques (1, 8, 9) have been described previously. The dye-fade times and the times of firm gel formation were determined for each run. Aliquots (50 µl) were withdrawn from the solutions period-ically and the concentration of remaining acrylamide monomer was

129

determined by gas-liquid chromatography. The concentrations of
reactants were those of Davis (1) as commonly used for formation
of large-pore polyacrylamide gels (10).

RESULTS AND DISCUSSION

The series of riboflavin-sensitized photopolymerizations
studied (Tables 1 and 2) show that many additives used in polyacrylamide gel formation have a pronounced effect on dye-fade time,
polymerization rate, and gel time at the low concentrations of
acrylamide-BIS normally used. Oster observed (6, 7) that riboflavin-sensitized photopolymerization of acrylamide is kinetically
dependent on the square of the monomer concentration, and in
previous studies (8) we found that 1-5% concentrations of acrylamide do not polymerize after 4 hr irradiation unless accelerators
or other additives are introduced into the solution, probably due
to oxygen inhibition of the slow polymerization rate. Furthermore,
Davis (1) and Hedrick and Smith (5) noted that solutions containing
less than 2% acrylamide-0.5% BIS do not photopolymerize to polyacrylamide gels, and use of low monomer concentrations tends to
magnify the effect of individual additives on photopolymerization.
Although these photopolymerizations are usually carried out with
exclusion of oxygen, complete and rigorous oxygen exclusion is
undesirable, since oxygen is involved in the photoinitiation system
(6-8). The effect of hydrogen peroxide formed during photoinitiation and polymer photografting (8, 9) on proteins during
photopolymerization should be considered, since they may introduce
artifacts which may be detectable during protein separations.

When additives used in formation of polyacrylamide gels from
riboflavin-acrylamide-BIS solutions are added individually, most
are found to effect the rate of gel formation (Table 1). When no
additives are present (Run 1), no photopolymerization occurs.
Rapid dye-fade times and photopolymerizations to gels are noted
when low concentrations of accelerators (Runs 2, 3, 5) are added to
solution; however, high triethanolamine (TEOA) concentrations
(Run 4) are not effective, due to the basic pH ($\approx$11) of the
resulting solution. Triethanolamine hydrochloride (TEOA·HCl) (Run 5)
in addition to the common accelerators N,N,N',N'-tetramethylethylene-
diamine (TEMED) and TEOA (Runs 2, 3) all effectively accelerate
photopolymerization in the absence of TRIS-HCl buffer.

Gel formation is not observed in the presence of TRIS (Run 6)
or dilute HCl (Run 7), although slow photopolymerization is found
in the presence of TRIS; however, in the presence of buffer
concentrations of TRIS-HCl (Run 8), rapid gel formation is observed
without addition of accelerators. Catalytic quantities of TRIS·HCl
(Run 9) cause photopolymerization, but gel formation is not observed.
It is evident that the presence of TRIS-HCl buffer is sufficient
for initiation of photopolymerization and subsequent gel formation

Table 1

The Effect of Additives on Riboflavin-Sensitized Photopolymerizations of Acrylamide-N,N'-Methylenebisacrylamide (BIS) Solutions[a]

Run No.	Additive(s) Name[b]	Conc. (g/100 ml)	Polymerization %	Polymerization Time (min)	Gel time (min)	Dye-fade time (min)
1	--	--	0	120	--	>60
2	TEMED	0.05	47	5	15	3
3	TEOA	0.05	--	--	10	2
4	TEOA	0.50	58	10	--	5
5	TEOA·HCl	0.06	--	--	5	1
6	TRIS	0.75	32	120	--	5
7	0.06 N HCl	--	0	120	--	>60
8	TRIS-0.06 N HCl	0.75	48	10	17	15
9	TRIS·HCl	0.06	50	120	--	>60
10	THPC	0.5	69	15	30	1
11	Sucrose	20.0	0	120	--	>60
12	Glycine	3.0	38	15	22	17
13	Glycine-NaCl	3.0-0.3	30	15	20	15

[a] Conc. of reactants - [sodium riboflavin-5'-phosphate] = 0.5 mg/100 ml, [acrylamide] = 2.5 g/100 ml, [BIS] = 0.625 g/100 ml.

[b] TEMED = N,N,N',N'-Tetramethylethylenediamine, TEOA-triethanolamine, TRIS-Tris(hydroxymethyl)aminomethane, THPC-Tetrakis(hydroxymethyl)phosphonium chloride, TEOA·HCl-TEOA hydrochloride, TRIS·HCl-TRIS hydrochloride.

Table 2

The Effect of Accelerators and Proteins on Riboflavin-Sensitized Photopolymerizations of Acrylamide-BIS Solutions in TRIS-HCl Buffer[a]

Run No.	Additive(s) Name[b]	Additive(s) Conc. (g/100 ml)	Polymerization %	Polymerization Time (min)	Gel time (min)	Dye-fade time (min)
14	--	--	25	10	18	12
15	TEMED	0.05	--	--	2	1
16	TEMED[c]	0.05	--	--	5	2
17	TEOA	0.05	5	5	15[d]	2
18	BSA	0.10	42	15	35	20
19	Lysozyme	0.10	40	10	40	17
20	BSA-TEMED	0.10-0.05	--	--	8	1
21	Lysozyme-TEMED	0.10-0.05	--	--	20	1

[a] Conc. of reactants - as in Table 1, except [TRIS] = 0.75 g/100 ml (0.06 $\underline{N}$) in 0.06 $\underline{N}$ HCl and [sucrose] = 20.0 g/100 ml.

[b] As in Table 1, except BSA - Bovine Serum Albumin.

[c] Oxygen excluded from run.

[d] Loose gel formed.

without added accelerator.

Tetrakis(hydroxymethyl)phosphonium chloride, an oxygen scavenger and reducing agent, causes photopolymerization and gel formation (Run 10) at 30 min if present in high concentrations and might be useful as an accelerator in special cases where nearly complete oxygen exclusion is desirable. No polymerization is found in the presence of sucrose (Run 11), a viscosity improver which gives more reproducible and rapid gel formation (1); however, glycine and glycine-NaCl, trailing ion sources (Runs 12, 13), cause rapid photopolymerization and gelation after an induction period. The above data show that satisfactory polyacrylamide gel formation is possible in the presence of individual additives. Therefore, it is possible to exclude some normally used additives from a particular system if such exclusion is desirable.

Solutions normally used for formation of electrophoretic gels contain TRIS-HCl buffer and sucrose in addition to riboflavin and monomers. Further addition of accelerators and proteins effect the rate of gel formation and dye-fade time (Table 2) of these solutions. Gel formation occurs in 18 min with a 12 min dye-fade time in the absence of accelerators or protein (Run 14), while addition of TEMED shortens the dye-fade time and gel time (Run 15) even when oxygen is excluded (Run 16). The short irradiation times necessary for gel formation suggest that the longer irradiation times used (1-5) are often unnecessary for complete polymerization. As Davis has observed (1), TEMED is more effective than TEOA (Run 17) which forms only a loose gel under our reaction conditions.

Proteins are observed to slow or inhibit gel formation (1). Bovine serum albumin (BSA) (Run 18) and lysozyme (Run 19) both slow gel formation when introduced into solutions; however, addition of TEMED shortens the dye-fade time and gel time. TEMED accelerator serves a most useful function by shortening the gel formation time, particularly when gels are formed in the presence of proteins, as is commonly the case.

Generally, additives which lower the dye-fade and gel times are mild reducing agents present in low concentrations, which are capable of donation of hydrogens. Recently, Gaylord (11) has shown the importance of donor-acceptor interactions in photoinitiated polymerizations, and this approach has been used to explain the behavior of numerous photopolymerizations found in the literature. It is quite possible that added hydrogen-donating, mild reducing additives complex with riboflavin and/or monomers through donor-acceptor interactions, to form species which possess sufficient photoreactivity to initiate photopolymerization successfully. Since initiation through donor-acceptor interactions seems probable, study of such interactions should be useful in development of new photoinitiated gel systems.

In summary, solution additives have a marked effect on photoinitiation and polymerization of dilute polyacrylamide-BIS
solutions to give gels suitable for bioanalytical separations.
Furthermore, it is apparent that some additives normally used are
unnecessary for gel formation in many cases, and care must be
exercised in selecting components for a given solution to assure
gel formation without introduction of unneeded components and
possible gel artifacts.

REFERENCES

1. Davis, B. J., Ann. N.Y. Acad. Sci. 121, 404 (1964) and refer-
    ences therein.

2. Antoine, B., Science 138, 977 (1962).

3. Mengoli, H. F. and Watne, A. L., Nature 212, 481 (1966).

4. Jolley, W. B., Allen, H. W. and Griffith, O. M., Anal. Biochem.
    21, 454 (1967).

5. Hedrick, J. L. and Smith A. J., Arch. Biochem. Biophys. 126,
    155 (1968).

6. Oster, G. K., Oster, G. and Prati, G., J. Am. Chem. Soc. 79,
    595 (1957).

7. Oster, G., Bellin, J. S. and Holstrom, B., Experientia 18,
    249 (1962).

8. Needles, H. L., J. Polymer Sci. B5, 595 (1967).

9. Needles, H. L., J. Appl. Polymer Sci. 12, 1557 (1968).

10. "Chemical Formulations for Disc Electrophoresis," Canal
    Industrial Corporation, Bethesda, Maryland (1963).

11. Gaylord, N. G., Polymer Preprints, Am. Chem. Soc. 10, 277
    (1969) and references therein.

POLYPERFLUOROBUTADIENE. V. PHOTOPOLYMERIZATION OF

PERFLUOROBUTADIENE

Madeline S. Toy

Douglas Advanced Research Laboratories
McDonnell Douglas Corporation
Huntington Beach, California 92547

Despite the ease of polymerization of tetrafluoroethylene, the homopolymerization of other perfluoroolefins and perfluoro-dienes is difficult to accomplish (1). The homopolymerization of hexafluoropropene to high polymer was brought about under extreme conditions, at a temperature of 210°C and using pressures in the range of 3000-5000 atm. (2). A heavy oil was reported by Roberts, when a mixture of bis(trifluoromethyl)peroxide and hexafluoropro-pene was irradiated by Hg 2537 A° radiation (3). Wall and his coworkers attacked the polymerization resistance problem using γ-ray initiation and high pressure (up to 20,000 atm.). High polymers were obtained from terminally unsaturated perfluoro-pentadiene (4), perfluorohexadiene, perfluoroheptadiene and per-fluorooctadiene (5). The polymerization of perfluorobutadiene was reported by Miller to give high polymer under very high pres-sure with oxygen and peroxide promotion (6). Recently Toy and her coworkers reported bulk polymerization of perfluorobutadiene under low pressure and temperature in the presence of free radical catalysts to give polyperfluoro-1,2- and 1,4-butadiene (7,8,9). In this paper, the photopolymerizations of perfluorobutadiene by γ-ray initiation, ultraviolet light and in the presence of CF_3OOCF_3 are described.

Discussion of Results

Although the ultraviolet spectrum of perfluorobutadiene (Figure 1a) shows the rapid absorption increase from 260 to 230 mμ to the total absorption below 230 mμ, photolysis of the monomer by ultraviolet light gave very poor yield of the polymer (below 1 to 3%). Radiation induced by γ-rays increased yield to 50% conversion, but similar waxy product of molecular weight between

135

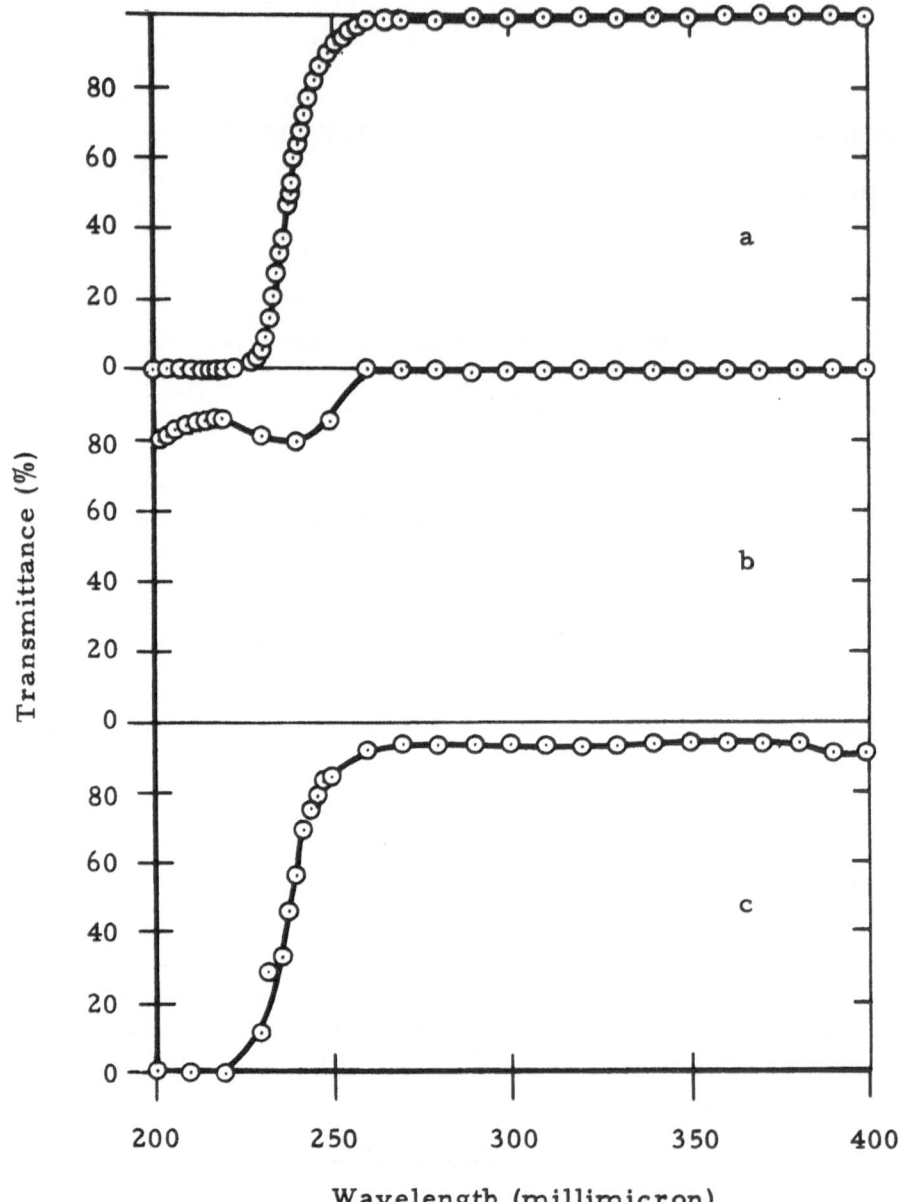

Figure 1.   Ultraviolet Spectra of (a) Perfluorobutadiene
            Vapor at 30 mm Hg; (b) CF_3OOCF_3 Vapor at
            30 mm Hg; (c) Mixed Vapor of Perfluoro-
            butadiene (40 mm Hg) and CF_3OOCF_3
            (20 mm Hg) in 1 cm Cell at Room Temperature

1500 to 2500 was obtained. The infrared spectra (Figures 2 and 3)
show a strong broad band between 7.5 and 9.1μ indicating C-F
absorption. The 5.6μ band identifies as pendant perfluorovinyl
groups and 5.8μ band as perfluorovinylene groups. The product is
thus polyperfluoro-1,2-and 1,4-butadiene.

A substantial increase of yields and molecular weights of
polyperfluorobutadiene were obtained, when photolysis of perfluorobutadiene by ultraviolet light was carried out in the presence of
free radical catalysts such as benzoyl peroxide, t-butyl hydroperoxide and bis(trifluoromethyl)peroxide. The results of photopolymerization in the presence of CF_3OOCF_3 were summarized in
Table I. The infrared spectra show that the ratio of peak intensity at 5.6μ indicating 1,2-polymer to 5.8μ indicating 1,4-poly-
mer is higher for polyperfluorobutadiene initiated by photolysis
in the presence of CF_3OOCF_3 as catalyst than without radiation.
The X-ray diffraction patterns of the two homopolymers were also
different as shown in Table II.

The first step of chain initiation is the breaking of the
relatively weak O-O bond to give $CF_3O\cdot$ radicals,

$$CF_3OOCF_3 \rightleftharpoons 2CF_3O\cdot$$

after which a chain mechanism takes place. Table III shows an
increase of catalyst concentration following an increase in yield
of the homopolymer and a decrease of softening point. A preliminary rate of polymerization curve using CF_3OOCF_3 as catalyst
under photolysis is shown in Figure 4.

Experimental

Monomer

Perfluorobutadiene was obtained from the Peninsular Chem-
Research Co., and was shown by Mitsch and Neuvar to be 98-99%
pure by vapor phase chromatography (10). The commercial monomer
showed three groups of lines for its F^{19}NMR spectrum as reported
by Toy and Lawson (7), and identical infrared spectrum reported
by Weiblen (11). The ultraviolet absorption spectrum is shown
in Figure 1a.

Photopolymerization Induced by γ-rays

Perfluorobutadiene (2 ml) was vacuum sealed in a Pyrex
ampoule (3-1/2" x 1/2" o.d.) and exposed to 9.6 kilocuries of
γ-rays from a cobalt-60 source for 5 days at ambient temperature
to give 50% conversion of soft wax. The solid was soluble in
hot hexafluorobenzene and 3M Brand Fluorochemical FC-43 and

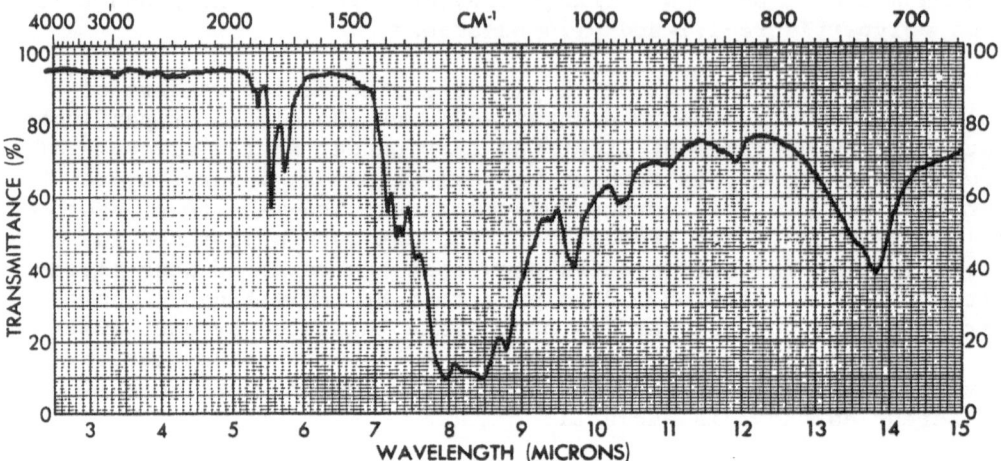

Figure 2.   Infrared Spectrum of Polyperfluorobutadiene
Initiated by γ-Rays

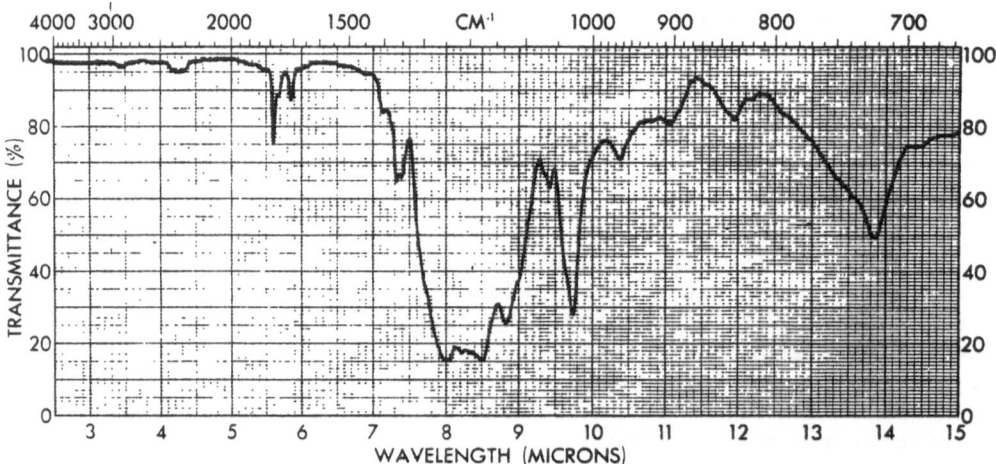

Figure 3.   Infrared Spectrum of Polyperfluorobutadiene
Initiated by CF_3OOCF_3 Under Ultraviolet
Irradiation (Batch No. 5)

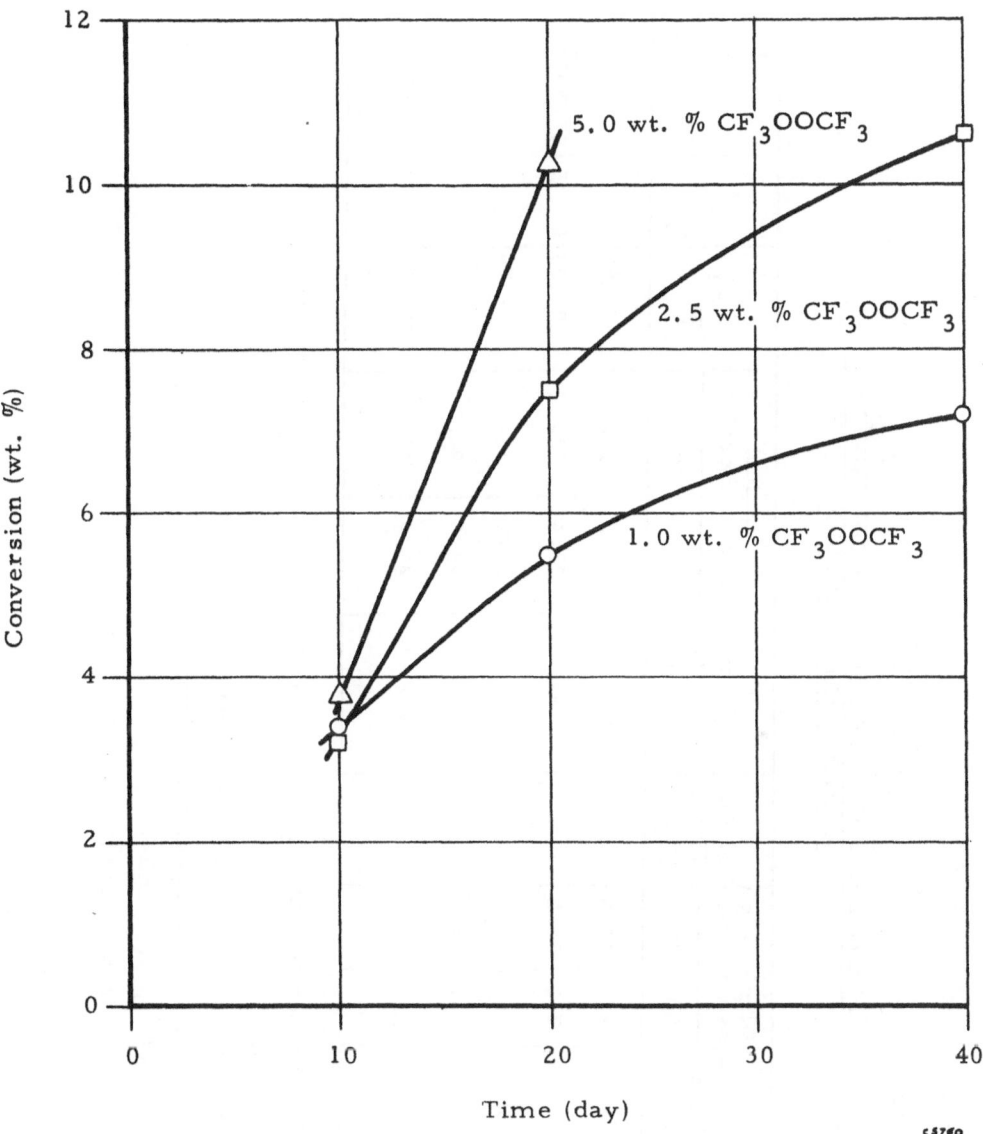

Figure 4.   Polymerization of Perfluorobutadiene Initiated by
CF$_3$OOCF$_3$ Under Ultraviolet Irradiation

TABLE I

POLYMERIZATION OF PERFLUOROBUTADIENE AT AMBIENT TEMPERATURE

Batch No.	Perfluoro-butadiene (g)	CF₃OOCF₃ as Catalyst (g)	(%)	Polym. Time (mo)	Exposure to 275-watt Sun Lamp[a]	Product Yield Conversion (g)	(%)	T_m[b] (°C)	$[\eta]$[c] (dl/g)	M_n[d]	DP[e]
1[f]	8.8	1.1	12.5	8.0	No	1.0	11.4	157-160	—		
2	44.9	7.1	15.8	0.9	Yes	23.4	52.1	80-105	0.045	10,300	63
3	43.4	3.9	9.0	0.9	Yes	17.9	41.2	75-100			
4	45.6	6.3	13.8	3.8	Yes	43.0	94.5	<50			
5[f']	93.6	2.4	2.6	3.6	Yes	43.9	46.9	90-100	0.032	7,800	48

(a) From General Electric Company at 9 inches distance from quartz reaction vessel.

(b) In sealed capillary tube.

(c) In hexafluorobenzene at 30.1°C. Sign "⌐" designates insolubility at 30.1°C.

(d) Refer to Reference 7 for $[\eta]$ vs. mole. wt. relationship.

(e) Approximate number of repeating units.

(f) See Table II for X-ray diffraction data.

(f') See Figure 3 for infrared spectrum and (f).

partially soluble in hot 3M's FC-75. The infrared spectrum of
the solid is shown in Figure 2.

Photolysis by Ultraviolet Light

Perfluorobutadiene (2 ml) was vacuum sealed in a quartz ampoule (3-1/2" x 1/2"o.d.) and photolyzed by a high mercury arc
lamp (A-H6) between 0° to 5°C for one hour. The liquid was condensed into a Pyrex reaction tube after photolysis, sealed under
vacuum and placed in a 60°C bath. The very small amount of residual solid appearing as a white ring around the quartz tube was
insoluble in hot hexafluorobenzene. The infrared spectrum shows
the characteristic broad band of C-F as well as the 5.8µ peak for
perfluorovinylene groups, but the intensity of 5.6µ peak for
perfluorovinyl groups was extremely weak.

The unreacted monomer, which was sealed into a Pyrex reaction
tube and placed in a 60° bath, polymerized slowly. After two
months, a thin layer of polymeric solid about 1% conversion
covered the inside wall of the reaction tube. The infrared spectrum of the waxy solid is similar to Figure 2.

Another sealed quartz reaction tube containing the monomer
as described above was placed 9 inches from a 275-watt sun lamp
for a week at ambient temperature to give less than 1% conversion
of white waxy solid. Its infrared spectrum is also similar to
Figure 2.

Radiation Induced by Ultraviolet in the Presence of
Bis(trifluoromethyl)peroxide as Catalyst

The ultraviolet spectra of perfluorobutadiene, bis(trifluoromethyl)peroxide (Peninsular ChemResearch) and the mixed vapor of
the monomer and the catalyst are shown in Figure 1.

A series of bulk polymerizations was carried out by conden-
sing monomer into the pressure bottles (Pyrex or quartz) with
metal caps and two-way valves. The clear homogeneous solutions
in the pressure bottles under autogenous pressure (i.e., about
2 atm) were subjected to vigorous stirring (except Batch No. 1 of
Table I) at ambient temperature. The source of ultraviolet irradiation was a General Electric 275-watt sun lamp 9 inches from the
quartz reaction vessels. The unreacted monomer at the end of
each experiment was discharged into another container through the
vacuum system after cooling the container. The white resins
were dried at 50°C under reduced pressure overnight. The results
of polyperfluorobutadiene preparations are summarized in Table I.

The X-ray diffraction patterns (Table II) were prepared by

TABLE II

X–RAY DIFFRACTION PATTERNS
OF POLYPERFLUOROBUTADIENE

| Line | Batch No. 1[a] | | Batch No. 2[a] (Under Ultraviolet) | |
	d-Spacing, $\overset{o}{A}$	Intensity	d-Spacing, $\overset{o}{A}$	Intensity
1	5.20	Strong	3.55	Very strong
2	4.40	Medium	3.15	Very weak
3	4.10	Weak	2.90	Strong
4	3.80	Very weak	2.25	Very weak
5	3.55	Very weak	1.73	Weak
6	2.65	Very weak	1.52	Weak
7	2.32	Very weak	1.00	Weak

(a) Refer to Table I

TABLE III

EFFECT OF CHANGING MONOMER-CATALYST RATIO ON YIELDS AND PHYSICAL PROPERTIES OF POLYPERFLUOROBUTADIENE[a]

Monomer, C_4F_6 (mmole)	Catalyst CF_3OOCF_3 (mmole)	Conversion[b] (%)	T_m[c] ($^\circ$C)
93	4.4	10.3	149–161
93	2.2	7.5	151–156
93	0.88	5.5	160–167

(a) Temperature 35°C, 20 days, under autogenous pressure about 2 atm, and exposure to a 275-watt sunlamp at 1-foot from quartz reaction tube.

(b) Based on grams of polymer recovered.

(c) From DTA measurements by differential scanning calorimeter of Perkin-Elmer Model DSC-1B.

the use of CuKα radiation (λ = 1.542A°) in a camera of 57.3 mm radius. The powder, prepared by grinding in an agate mortar at ambient temperature, was mounted on a quartz fiber about 0.1 mm in diameter, using Vaseline as a binder.

ACKNOWLEDGMENT

This work is partially sponsored by McDonnell Douglas Independent Research and Development funds, and partially sponsored by National Aeronautics and Space Administration under Contracts Nos. NAS7-603 and NAS7-723.

REFERENCES

1. R. M. Adams and F. A. Bovey, J. Polymer Sci., 9, 481 (1952).

2. H. S. Elenterio, U. S. Patent 2,958,685 (E. I. DuPont de Nemours & Co.),1960.

3. H. L. Roberts, J. Chem. Soc., 4538 (1964).

4. J. E. Fearn, D. W. Brown and L. A. Wall, J. Polymer Sci., Part A-1, 4, 131 (1966).

5. L. A. Wall, Am. Chem. Soc. Polymer Preprints, 7 (2), 1112 (1966).

6. W. T. Miller, "Preparation, Properties and Technology of Fluorine and Organic Fluoro Compounds," ed. by C. Slesser and S. R. Schram, McGraw-Hill, N.Y., 1951, pp. 624-626 and p. 604.

7. M. S. Toy and D. D. Lawson, J. Polymer Sci., Part B, 6, 639 (1968).

8. M. S. Toy and D. D. Lawson, Am. Chem. Soc, Polymer Preprints, 9 (2), 1671 (1968).

9. M. S. Toy and J. M. Newman, J. Polymer Sci., Part A-1, 6, 2333 (1969).

10. R. A. Mitsch and E. W. Neuvar, J. Phys. Chem., 70 (2), 546, (1966).

11. D. G. Weiblen, Fluorine Chemistry, Vol. 2, ed. by J. H. Simons, Academic Press, N.Y., 1954, p. 478.

LIGHT SCATTERING SPECTROSCOPY AS A TOOL FOR STUDYING MACROMOLECULAR DYNAMICS AND CHEMICAL KINETICS*

R. Pecora[+]

Department of Chemistry, Stanford University

Stanford, California 94305

I. Introduction

Light scattering has been used for about the past 25 years to determine such static properties of polymers in solution as molecular weights, solution virial coefficients, molecular shapes and characteristic dimensions, molecular optical anistropies and molecular weight and size distributions of polydisperse samples.[1] In all of these cases the measured quantity is the frequency integrated intensity of scattered light as a function of scattering angle. In experiments of this type, no attempt is made to measure the spectral distribution of the scattered light. Since the width of the scattered spectral distribution is usually very small, the tools to perform this additional analysis were not available. Recently, however, with the development of lasers and associated techniques such measurements have become commonplace. This new field, light scattering spectroscopy, extends the light scattering technique to the measurement of dynamic quantities, such as translational and rotational diffusion coefficients, intramolecular relaxation times and chemical reaction rate constants.

Light scattering spectroscopy is one of the new tools developed in the past few years for studying dynamics of fluids and polymers. The application to polymers was first suggested by the author in 1962[2]; the first experiment was performed on polystyrene latex sphere suspensions in 1964 by H. Z. Cummins, N. Knable and Y. Yeh[3]. Since these first attempts several

*Work supported by the National Science Foundation.

[+]Alfred P. Sloan Foundation Fellow.

experimental[4-9] and theoretical[10-16] articles on this subject have
been published. There are now at least twenty laboratories making
a major effort to develop this field, although most of them have
not yet reached the publishing stage. In this article, we give a
brief review of the theory of light scattering spectra of macromolecular solutions. We hope also to convey an understanding of
why this new field has aroused so much interest.

II. Formal Theory

Consider a polymer solution upon which is incident a plane
polarized light wave. We develop the theory here for a special but
rather widely used scattering geometry. The incident wave propa-
gates in the x-y plane with propagation vector $\underset{\sim}{k}_0$ and electric field
vector $\underset{\sim}{E}_0$ polarized along the direction of the z-axis. The scattered wave propagates in a direction parallel to the positive x-
axis with propagation vector $\underset{\sim}{k}_f$. The scattering angle θ is defined
as the angle between $\underset{\sim}{k}_0$ and $\underset{\sim}{k}_f$ and the scattering vector $\underset{\sim}{K} \equiv \underset{\sim}{k}_0 - \underset{\sim}{k}_f$.
We note that if frequency changes in the scattering are small, as
is usually the case, that the length of $\underset{\sim}{K}$ is given by

$$K = \frac{4\pi}{\lambda_0} \sin\frac{\theta}{2} \qquad\qquad (1)$$

where λ_0 is the wavelength of the incident light in the medium.

The scattering medium is divided in N scattering "segments"
Segment (i) has polarizability tensor $\alpha^{(1)}(\Omega)$ in a laboratory fixed
coordinate system. The Euler angles α, β, γ through which the
laboratory fixed coordinate coordinate system must be rotated in
order to coincide with a molecule-fixed coordinate system are
represented by the single symbol Ω. For the given geometry, it has
been shown elsewhere[12], that the scattered light intensity at a
distance R, far removed from the scattering medium at a frequency
change ω is given by

$$I(K,\omega) = I_{VV}(K,\omega) + I_{HV}(K,\omega) \qquad\qquad (2)$$

where I_{VV} the intensity of light scattered with vertical polarization is proportional to

$$I_{VV}(K,\omega) = A\frac{1}{2\pi}\int dt \int_V d^3r \, \exp i(\underset{\sim}{K}\cdot\underset{\sim}{r} - \omega t)$$

$$\int_V d^3r \langle \alpha_{zz}(\underset{\sim}{r} + \underset{\sim}{r}',t)\alpha_{zz}(r',0)\rangle \qquad\qquad (3)$$

The intensity of light scattered with polarization in the x-y
plane I_{Hv} is proportional to

$$I_{HV}(K,\omega = A\frac{1}{2\pi}\int dt \int_V d^3r \ \exp \ i(\underline{K}\cdot\underline{r} - \omega t)$$

$$\int_V d^3r'\langle\alpha_{yz}(\underline{r} + \underline{r}',t)\alpha_{yz}(\underline{r}',0)\rangle \quad (4)$$

where $\alpha_{zz}(\underline{r},t)$ and $\alpha_{yz}(\underline{r},t)$ are, respectively, the zz and yz elements of the <u>fluid</u> polarizability tensor at point $\underline{r}$ and time t, and A is a constant which depends on k_f, R, E_0, the average dielectric constant of the medium ϵ, and the scattering volume V,

$$A = \frac{k_f^4|E_0|^2}{16\pi^4R^2\epsilon^2V} \quad (5)$$

The angular brackets in Eqs. (3) and (4) denote statistical averages and the spatial integrations are over the volume V.

We now <u>assume</u> that the polarizability of the fluid may be expressed in terms of the segmental polarizabilities defined above,

$$\alpha_{zz}(\underline{r},t) = \sum_{i=1}^{N} \alpha_{yz}^{(i)}(\Omega,t)\delta(\underline{r}_i(t) - \underline{r})$$

and   (6)

$$\alpha_{yz}(\underline{r},t) = \sum_{i=1}^{N} \alpha_{yz}^{(i)}(\Omega,t)\delta(\underline{r}_i(t) - \underline{r})$$

where $\underline{r}_i(t)$ is the position of segment (i) at time t and $\delta(\underline{r}_i(t) - \underline{r})$ is the Dirac delta function. Eqs. (6) are an important assumption in the theoretical development to follow. They neglect fluctuations of polarizability due to mutual interactions of segments.

Substituting Eqs. (6) into Eqs. (3) and (4), we obtain, after defining some new symbols for convenience,

$$C_{zz}(K,\omega) \equiv \frac{I_{vv}}{VA} = \equiv \frac{1}{2\pi}\int C_{zz}(K,t)e^{-i\omega t}dt$$

and   (7)

$$C_{yz}(K,\omega) \equiv \frac{I_{Hv}}{VA} = \frac{1}{2\pi}\int C_{yz}(K,t)e^{-i\omega t}dt$$

where

$$C_{zz}(K,t) = \frac{1}{V} \sum_{i,j} \langle\alpha_{zz}^{(i)}(\Omega,t)\alpha_{zz}^{(j)}(\Omega_0,0)\exp\left\{i\underline{K}\cdot(\underline{r}_i(t)\right.$$

$$\left. - \underline{r}_j(0))\right\}\rangle \quad (8)$$

and

$$C_{yz}(K,t) = \frac{1}{V} \sum_{i,j} \langle\alpha_{yz}^{(i)}(\Omega,t)\alpha_{yz}^{(j)}(\Omega_0,0)\exp\left\{i\underline{K}\cdot(\underline{r}_i(t)\right.$$

$$\left. - \underline{r}_j(0))\right\}\rangle. \quad (9)$$

The frequency integrated spectra that have been so routinely used in polymer studies are given by $C_{zz}(K,0)$ and $Cyz(K,0)$[1].

Although in some cases it might be necessary to use the more general forms Eqs. (4) and (5), we use the relatively simple expressions Eqs. (8) and (9) in the theory developed in this article. Equations (8) and (9) should be best for uncharged, non-polar segments in nonpolar, non-electrolytic solvents, although little is definitely known about the validity of the approximation.

In order to proceed further, we must express the laboratory system segmental polarizabilities in terms of those in a molecule fixed system and some functions of the molecular orientations. This is most easily done using a spherical tensor formulation. A set of spherical tensor components may be defined as

$$\alpha_I = \frac{1}{3}(\alpha_{xx} + \alpha_{yy} + \alpha_{zz})$$

and

$$\alpha_{-2} = \frac{1}{2}(\alpha_{xx} - \alpha_{yy}) - i\alpha_{xy}$$

$$\alpha_{-1} = (\alpha_{zx} - \alpha_{zy})$$

$$\alpha_0 = \frac{2}{\sqrt{6}}(\alpha_{zz} - \frac{1}{2}(\alpha_{xx} + \alpha_{yy})) \tag{10}$$

$$\alpha_{+1} = -(\alpha_{-1})^*$$

$$\alpha_{+2} = (\alpha_{-2})^*$$

With these definitions of the spherical components the laboratory-fixed components $\alpha_n(\Omega)$ are related to the molecule-fixed components α_n by

$$\alpha_n(\Omega) = \left(\frac{8\pi^2}{5}\right)^{\frac{1}{2}} \sum_{n'=-2}^{+2} D^2_{n,n'}(\Omega)\alpha_{n'} \tag{11}$$

and the α_I are invariant to rotation. The $D^2_{n,n'}(\Omega)$ are the Wigner rotation matrices. The conventions on these matrices are those used by Valiev et al, except that Valiev's $D^2_{n,n'}$ must be multiplied by $[\frac{5}{8\pi^2}]^{\frac{1}{2}}$ to give our matrices[17].

Using Eq. (10), Eqs. (18) and (9) become

$$VC_{zz}(K,t) = \sum_{i,j} \alpha_I^{(i)} \alpha_I^{(j)} \langle \exp\{i\underset{\sim}{K} \cdot (\underset{\sim}{r_i}(t) - \underset{\sim}{r_j}(0))\}\rangle$$

$$+ \sum_{i,j} \frac{\sqrt{6}}{3}\{\langle \alpha_0^{(i)}(\Omega,t)\alpha_I^{(j)}$$

$$+ \alpha_I^{(i)}\alpha_0^{(j)}(\Omega,t)\}\exp\{i\underset{\sim}{K} \cdot (\underset{\sim}{r_i}(t) - \underset{\sim}{r_j}(0))\}\rangle$$

$$+ \frac{2}{3} \sum_{i,j} \langle \alpha_0^{(i)}(\Omega,t)\alpha_0^{(j)}(\Omega_0,0)$$

$$\times \exp\{i\underset{\sim}{K} \cdot (\underset{\sim}{r_i}(t) - \underset{\sim}{r_j}(0))\}\rangle \qquad (12)$$

$$C_{yz}(K,t) = \frac{1}{4V} \sum_{i,j}^{N} \langle \{\alpha_1^{(i)^*}(\Omega,t) + \alpha_{-1}^{(i)^*}(\Omega,t)\} \times$$

$$\{\alpha_1^{(j)}(\Omega_0,0) + \alpha_{-1}^{(j)}(\Omega_0,0)\} \times$$

$$\exp\{i\underset{\sim}{K} \cdot (\underset{\sim}{r_i}(t) - \underset{\sim}{r_j}(0))\}\rangle \qquad (13)$$

Equation (12) indicates that in general the C_{zz} component of the scattering may be divided into an "isotropic" part proportional to the product of isotropic polarizability components $\alpha_I^{(i)}\alpha_I^{(j)}$, an "interaction" part proportional to $\alpha_I^{(i)}\alpha_0^{(j)} + \alpha_0^{(i)}\alpha_I^{(j)}$ and a purely "anisotropic" part proportional to $\alpha_0^{(i)}\alpha_0^{(j)}$. When the segments are optically isotropic and the polarizability principal axes coinside with those for segmental motion, only the isotropic term remains. In this case, the C_{yz} component which has no dependence on the $\alpha_I^{(i)}$ at all vanishes.

III. Application to Dilute Macromolecular Solutions

(A) General Equations

In this section, we specialize the theory developed above to dilute polymer solutions. Theories for more concentrated polymer solutions or polymer melts have not yet been put forward. As an illustration of a simple system that may be studied by the light scattering method, we develop the theory for a dilute solution of identical rigid rod molecules undergoing independent translational and rotational Brownian motion. We first write out the general equations for dilute polymer solutions.

Consider a dilute solution of N_m polymer molecules with each molecule divided into N_s identical segments. Any interference between light waves scattered from different molecules will be ignored. This means that terms in Eqs. (12) and (13) in which segments (i) and (j) belong to different molecules will be set equal to zero. Furthermore the scattering from solvent molecules is considered to be neglibible.

We may also refer the position of each segment to the position of the center of mass $\underset{\sim}{R}_{cm}$ of the molecule to which it belongs. Thus

$$\underset{\sim}{r}_i(t) = \underset{\sim}{R}_{cm}(t) + \underset{\sim}{b}_i(t) \tag{14}$$

We define

$$\underset{\sim}{R}_t = \underset{\sim}{R}_{cm}(t) - \underset{\sim}{R}_{cm}(0) \tag{15}$$

From Eqs. (11) - (15), we find

$$
\begin{aligned}
C_{zz}(K,t) = {}& \rho\alpha_I^2 \sum_{i,j}^{N_s} \langle \exp(i\underset{\sim}{K}\cdot\underset{\sim}{R}_t)\exp\{i\underset{\sim}{K}\cdot(\underset{\sim}{b}_i(t) - \underset{\sim}{b}_j(0))\}\rangle \\
& + \rho\alpha_I \frac{\sqrt{6}}{3}\left(\frac{8\pi^2}{5}\right)^{\frac{1}{2}} \sum_{i,j}^{N_s}\sum_n \langle[\alpha_n^* D_{0,n}^{2*}(\Omega_i(t)) \\
& + \alpha_n D_{0,n}^2(\Omega_j(0))] \\
& \quad \exp(i\underset{\sim}{K}\cdot\underset{\sim}{R}_t)\exp\{i\underset{\sim}{K}\cdot(\underset{\sim}{b}_i(t) - \underset{\sim}{b}_j(0))\}\rangle \\
& + \rho\frac{2}{3}\left(\frac{8\pi^2}{5}\right) \sum_{i,j}^{N_s}\sum_{n,n'} \langle \alpha_n^*\alpha_{n'} D_{0,n}^{2*}(\Omega_i(t)) \times \\
& \quad D_{0,n'}^2(\Omega_j(0))\exp(i\underset{\sim}{K}\cdot\underset{\sim}{R}_t)\exp\{i\underset{\sim}{K}\cdot(\underset{\sim}{b}_i(t) - \underset{\sim}{b}_j(0))\}\rangle
\end{aligned} \tag{16}
$$

and

$$
\begin{aligned}
C_{yz}(t) = {}& \frac{2\pi^2\rho}{5} \sum_{i,j}^{N_s}\sum_{n,n'}\sum_{sgn} \langle \alpha_n^*\alpha_{n'}[D_{\pm1,n}^{*2}(\Omega_i(t) \times \\
& D_{\pm1,n'}^2(\Omega_j(0))\exp(i\underset{\sim}{K}\cdot\underset{\sim}{R}_t)\exp\{i\underset{\sim}{K}\cdot(\underset{\sim}{b}_i(t) - \underset{\sim}{b}_j\cdot(0))\}\rangle
\end{aligned} \tag{17}
$$

where $\rho = \frac{N_m}{V}$ and $\sum\limits_{sgn}$ means a sum over all combinations of $\pm$ signs.

We note here that we have used normalizing procedures for Eqs. (8), (9), (16) and (17) different from those used in reference 12. In this reference we expressed the results in terms of "average segment" α_I and α_n, and computed what is in the present

notation $\dfrac{I_{VV}}{N_s^2}$ and $\dfrac{I_{HV}}{N_s^2}$. The conventions used in the present treatment are probably the most convenient for relating the results to experiment.

Eqs. (16) and (17) give rather complicated expressions for the scattering. In order to evaluate the averages, we must know the structure and dynamics of the molecule including quantities not usually calculated in polymer hydrodynamics. For instance, effects such as the coupling of translational motion to intra-molecular motion might be important in light scattering spectra.

Rather than attempting to develop general techniques of calculation here, we treat a simple but rather important model -- the rigid rod molecule. This model illustrates the physical principles of the scattering and is applicable to a very wide class of molecules, e.g., helical polypeptides and tobacco mosaic virus.

(B) Rigid Rod Theory

The molecule is considered to consist of N_s identical scattering segments arranged along a line of total length L. The thickness of the rod is negligible compared to its length. We compute first the "isotropic" part of C_{zz} defined by

$$C_{zz}^{I}(K,t) \equiv \rho \alpha_{I}^{2} \sum_{i,j}^{N_s} \langle \exp(i\underset{\sim}{K} \cdot \underset{\sim}{R}_t)$$

$$\exp\left\{i\underset{\sim}{K} \cdot \left(\underset{\sim}{b}_i(t) - \underset{\sim}{b}_j(0)\right)\right\}\rangle \tag{18}$$

The dynamical assumptions are (1) rotation and translation may be considered independently; (2) the translational motion may be described by the translational diffusion equation,

$$\frac{\partial P_T(\underset{\sim}{R}_t,t)}{\partial t} = D\nabla^2 P_T(\underset{\sim}{R}_t,t) \tag{19}$$

where D is the average rotational diffusion coefficient of the rod, P_T the probability of finding the center of mass of the rod at R_t at time t and ∇^2 the Laplacian operator; (3) the rotational motion may be described by the rotational diffusion equation

$$\frac{\partial P_R(\Omega,t)}{\partial t} = \Theta \Delta P_R(\Omega,t) \tag{20}$$

where Θ is the rod rotational diffusion coefficient and $P_R(\Omega,t)$ is the probability of finding the long axis of the rod with orientation Ω at time t and Δ is the Laplacian operator on the surface of a unit sphere.

Equations (19) and (20) may be solved with the respective initial conditions

$$P_T(\underset{\sim}{R}_t, 0) = \delta(\underset{\sim}{R}_t) \tag{21}$$

and

$$P_R(\Omega, 0) = \delta(\Omega - \Omega_0) \tag{22}$$

The solutions are

$$P_T(R_t, t) = \frac{1}{(4\pi Dt)^{\frac{3}{2}}} \exp(-\frac{R_t^2}{4Dt}) \tag{23}$$

and

$$P_R(\Omega, \Omega_0, t) = \sum_{\ell, m} Y_\ell^m(\Omega) Y_\ell^{m*}(\Omega_0) \tag{24}$$

$$\exp\left\{-\Theta(\ell)(\ell + 1))t\right\}$$

where the $Y_\ell^m(\Omega)$ are the spherical harmonics.

From assumption (1), Eq. (18) may be written

$$c_{zz}^I(\kappa, t) = \rho\alpha_I^2 \langle \exp(i\underset{\sim}{\kappa}\cdot\underset{\sim}{R}_t) \rangle_T \langle \sum_{i,j}^{N_s} \exp\left\{i\underset{\sim}{\kappa}\cdot(\underset{\sim}{b}_i(t)\right.$$

$$\left. - \underset{\sim}{b}_j(0))\right\} \rangle_R \tag{25}$$

From assumption (2) and Eq. (23),

$$\langle \exp(i\underset{\sim}{\kappa}\cdot\underset{\sim}{R}_t) \rangle_T = \int P_T(\underset{\sim}{R}_t, t) \exp(i\underset{\sim}{\kappa}\cdot\underset{\sim}{R}_t) d^3R_t$$

$$= \exp(-\kappa^2 Dt) \tag{26}$$

Calculation of the rotational average is somewhat more complicated. We first note that some simplification results from the symmetry of the rod. We write

$$\langle \sum_{i,j}^{N_s} \exp\left\{i\underset{\sim}{\kappa}\cdot(\underset{\sim}{b}_i(t) - \underset{\sim}{b}_j(0))\right\} \rangle_R =$$

$$\langle \sum_{i=1}^{N_s} \exp(i\underset{\sim}{\kappa}\cdot\underset{\sim}{b}_i(t)) \sum_{j=1}^{N_s} \exp(-i\underset{\sim}{\kappa}\cdot\underset{\sim}{b}_j(0)) \rangle_R \tag{27}$$

Since all distances are measured from the center of the rod and the rod is symmetrical, it may be seen that for every $\underline{b}_j(t)$ pointing along the axis of the rod, there is a $\underline{b}_k(t)$ such that $\underline{b}_k(t) = -\underline{b}_j(t)$. If we add all terms for which this is true, we obtain

$$\left\langle \sum_{i,j}^{N_s} \exp\left\{i\underline{K}\cdot(\underline{b}_i(t) - \underline{b}_j(0)\right\}\right\rangle_R =$$

$$4\left\langle \sum_{i,j}^{\frac{N_s}{2}} \cos(\underline{K}\cdot\underline{b}_i(t))\cos(\underline{K}\cdot\underline{b}_j(0))\right\rangle_R \qquad (28)$$

where the summation is now over half the rod, i.e., only over those segments on the side of the rod that point along the positive rod axis direction.

Then, from assumptions (1) and (3),

$$4\left\langle \sum_{i,j}^{\frac{N_s}{2}} \cos(\underline{K}\cdot\underline{b}_i(t))\cos(\underline{K}\cdot\underline{b}_j(0))\right\rangle_R =$$

$$\qquad (29)$$

$$\frac{1}{\pi}\int\left(\sum_{i,j}^{\frac{N_s}{2}} \cos(\underline{K}\cdot\underline{b}_i(t))\cos(\underline{K}\cdot\underline{b}_j(0))\right)P_R(\Omega,\Omega_0,t)d\Omega d\Omega_0$$

Using the expansion

$$\cos(\underline{K}\cdot\underline{b}_i(t)) = 4\pi \sum_{\ell \text{ even},m} i^\ell Y_\ell^m(\Omega)Y_\ell^{m*}(\Omega_K)j_\ell(Kb_{it}) \qquad (30)$$

(where j_ℓ is the spherical Bessel function of order ℓ) and Eq. (24), we obtain from Eq. (29),

$$\left\langle \sum_{i,j}^{N_s} \exp\left\{i\underline{K}\cdot(\underline{b}_i(t) - \underline{b}_j(0))\right\}\right\rangle_R = 4\left\{\left|\sum_{i=1}^{\frac{N_s}{2}} j_0(Kb_i)\right|^2\right.$$

$$\left. + 5\left|\sum_{i=1}^{\frac{N_s}{2}} j_2(Kb_i)\right|^2 \exp(-6\Theta t) + \cdots\right\} \qquad (31)$$

Combining Eqs. (7), (25), (26) and (31), we obtain for the spectral distribution of this system

$$
c_{zz}^{I}(\kappa,\omega) = \frac{4}{\pi}\rho\alpha_{I}^{2}\Big\{\Big|\sum_{i}^{\frac{N_s}{2}} j_0(\kappa b_i)\Big|^2 \frac{\kappa^2 D}{\omega^2+(\kappa^2 D)^2} +
$$

$$
5\Big|\sum_{i=1}^{\frac{N_s}{2}} j_2(\kappa b_i)\Big|^2 \frac{(6\Theta + \kappa^2 D)}{\omega^2+(6\Theta+\kappa^2 D)^2} + \cdots\Big\} \qquad (32)
$$

The first term on the right hand side of Eq. (32) depends only on
the translational diffusion coefficient while the second term depends
on both the translational and the rotational diffusion coefficients.
Numerical calculation shows that the second term is significant only
for scattering from large molecules observed at large scattering angles.
If we define a dimensionless scattering parameter by

$$
\kappa L \equiv \frac{4\pi L}{\lambda}\sin\frac{\theta}{2} \qquad (33)
$$

we find that the second term is significant only for $KL \gtrsim 5$. Thus,
for a large molecule the second term may be "turned off" by observing the scattering at low angles. Under these conditions, a fit
of the experimental results to the remaining single Lorentzian of
Eq. (32) allows easy extraction of the translational diffusion coefficient. By observation of the scattering at large angles and a
fit of the spectrum to the full two Lorentzian from given in Eq. (32)
with use of the already determined D, the rotational diffusion coefficient may be extracted from the data. It may also be shown by
numerical calculation that higher order terms in Eq. (32) are
usually negligible. Details of the calculation and results are
given in reference 11. An experiment confirming this theory for
tobacco mosaic virus solutions has already been performed[6].

   The qualitative behavior of the isotropic component given
above for rigid rods is also applicable to molecules of other shapes
and dynamics. At low values of the scattering parameter, the spectrum consists only of a single Lorentzian line with half-width
dependent only on the translational diffusion coefficient. In order
for rotational diffusion coefficients or intramolecular relaxation
times to be observed, the "size" of the structure change associated
with the thermal motion must be large enough to be "seen" by the
light wave. For rigid uniform spheres, it is clear that no rotation will be seen. A detailed calculation has been performed for
the isotropic component for flexible coils[10].

   Calculation of the other terms appearing in Eqs. (16) and (17)
is more difficult. We consider here only the case where the intramolecular exponential terms may be set equal to 1. Thus, the

resulting expressions will ignore structure effects such as those described for $C_{zz}^{I}(\kappa,t)$, and the theory should be valid only for scattering from small molecules or large molecules observed at small angles.

For the rigid rod, we have $\alpha_{xx} = \alpha_{yy}$; $\alpha_{-2} = \alpha_{2} = \alpha_{-1} = \alpha_{+1} = 0$; $\alpha_0 = \frac{2}{\sqrt{6}}(\alpha_{zz} - \alpha_{yy})$ and $\alpha_I = \frac{1}{3}(\alpha_{zz} + 2\alpha_{yy})$. Equation (17) becomes

$$C_{yz}(\kappa,t) = \frac{2}{5}\pi^2 \rho N_s^2 \exp(-\kappa^2 Dt).$$

$$\alpha_0^2 \sum_{sgn} \langle D_{\pm1,0}^{2*}(\Omega)D_{\pm1,0}^{2}(\Omega_0) \rangle \qquad (34)$$

Noting that the expansion in Eq. (24) may be written

$$P_R(\Omega,\Omega_0,t) = \sum_{J,K,M} D_{K,0}^{J}(\Omega)D_{K,0}^{J*}(\Omega_0)\exp\left\{-\Theta J(J + 1)t\right\},$$

we multiply Eq. (34) by $\dfrac{P_R}{8\pi^2}$ and integrate over Ω and Ω_0 to obtain

$$C_{yz}(\kappa,t) = \frac{1}{10}\rho N_s^2 \alpha_0^2 \exp(-\kappa^2 Dt)\exp(-6\Theta t).$$

From Eq. (7), we obtain for the spectrum

$$C_{yz}(\kappa,\omega) = \frac{1}{15\pi}\rho N_s^2 (\alpha_{zz} - \alpha_{yy})^2 \cdot \frac{(\kappa^2 D + 6\Theta)}{\omega^2 + (\kappa^2 D + 6\Theta)^2} \qquad (35)$$

Notice that in Eq. (35) the rotational diffusion coefficient contributes to the scattering even for small molecules and for forward scattering. This "anisotropic" term contains a contribution from the translational diffusion coefficient. It represents the translational diffusion of the "optical anisotropy".

The terms other than $C_{zz}^{I}(\kappa,t)$ which contribute to $C_{zz}(\kappa,t)$ may be calculated in a similar way in the same approximation as given for the $C_{yz}(\kappa,t)$. We omit any discussion of them here, but discuss them in some detail in connection with the reaction rate theory given in the next section. An experiment measuring C_{yz} for dilute tobacco mosaic virus solutions has recently been performed by Wada et al[7].

IV. Chemically Reacting Systems

In this section we extend the light scattering theory to chemically reacting systems. By "chemically reacting system" we mean one in which there are dynamic **equilibria** connecting some of the species in solution. In particular, we consider the simple

case of a dilute solution of molecules of types 1 and 2 in an
inert solvent. Species 1 and 2 are in dynamic equilibrium,

$$1 \overset{k_{21}}{\underset{k_{12}}{\rightleftharpoons}} 2.$$

Contributions to the scattering from the solvent are, as before,
assumed to be negligible. The development here is similar to that
given in a previous work[16]. A more formal development using nonequilibrium thermodynamics has been given for the isotropic scattering by Blum and Salzburg[15].

    To simplify the development we ignore all macromolecular
structure factors and shape fluctuations in the chemically reacting
system, i.e., the exponentials in Eq. (16) and (17) containing
internal coordinates will be set equal to one. These factors are
easy to include for the isotropic scattering by combining the
methods of Section III above with the theory given here. We also
consider only the theory for cylindrically symmetric molecules.

    The following modification of Eq. (8) must be evaluated

$$C_{zz}(K,t) = \frac{1}{V} \sum_{i=1}^{N_m} \langle \alpha_{zz}^{(i)}(\Omega,t)\alpha_{zz}^{(i)}(\Omega_0,0)\exp(i\underset{\sim}{K}\cdot\underset{\sim}{R}_t)\rangle \qquad (36)$$

where $\alpha_{zz}^{(i)}$ is the zz polarizability of molecule (i), and $\underset{\sim}{R}_t$ is the
distance traveled by the molecular center of mass in the time
interval t.

    It is clear that if molecule (i) changes from, say, state 1
to state 2 in the time interval t, its polarizability components
and rotational and translational diffusion coefficients will be
changed. To take account of this transformation, we must evaluate
the four correlation functions corresponding to the four sets of
initial and final states for molecule (i). It is easy to see that
if ρ_1 and ρ_2 are the equilibrium concentrations of species 1 and
2, respectively, that

$$C_{zz}(K,t) = \sum_{\alpha,\gamma=1}^{2} \rho_\alpha \langle \alpha_{zz}^\gamma(\Omega,t)\alpha_{zz}^\alpha(\Omega_0,0)$$

$$\exp(i\underset{\sim}{K}\cdot\underset{\sim}{R}_t)\rangle_{\gamma\alpha} \qquad (37)$$

where α_{zz}^α is the zz polarizability of species α and the subscript
on the brackets means that the average is to be taken with the
molecule in initial state α and final state γ. The expression for
$C_{yz}(K,t)$ may be written in a similar way.

In order to evaluate the averages, we must extend the dynamical equations, Eqs. (19) and (20), to take account of the two species chemical equilibrium. It was shown elsewhere[18] that if we let $P_\alpha(R_t,\Omega,t)$ be the probability of observing the molecule in state α with position R_t and orientation Ω at time t, that

$$\frac{\partial P_1(R_t,\Omega,t)}{\partial t} = D_1 \nabla^2 P_1(R_t,\Omega,t) + \Theta_1 \Delta P_1(R_t,\Omega,t) +$$
$$k_{12}P_2(R_t,\Omega,t) - k_{21}P_1(R_t,\Omega,t)$$

$$\frac{\partial P_2(R_t,\Omega,t)}{\partial t} = D_2 \nabla^2 P_2(R_t,\Omega,t) + \Theta_2 \Delta P_2(R_t,\Omega,t) +$$
$$k_{21}P_1(R_t,\Omega,t) - k_{12}P_2(R_t,\Omega,t) \tag{38}$$

where $k_{\alpha\gamma}$ is the chemical rate constant which describes the disappearance of molecules from the state γ, and Θ_α and D_α are, respectively, the rotational and translational diffusion coefficients of species α.

The functions needed are the solutions of Eqs. (38) subject to the boundary conditions at time 0,

$$P_{\gamma\alpha}(R_t,\Omega,\Omega_0,0) = \delta_{\gamma\alpha}\delta(R_t)\delta(\Omega - \Omega_0) \tag{39}$$

$P_{\gamma\alpha}(R_t,\Omega,\Omega_0,t)$ may be interpreted as the probability that a molecule which is in state α with orientation Ω_0 and position 0 at time zero is in state γ with orientation Ω and position R_t at time t.

Eq. (37) may be rewritten in terms of the spatial Fourier transform of $P_{\gamma\alpha}$,

$$\widetilde{P}_{\gamma\alpha}(K,\Omega,\Omega_0,t) = \int d^3R_t \exp(iK\cdot R_t)P_{\gamma\alpha}(R_t,\Omega,\Omega_0,t) \tag{40}$$

$$C_{zz}(K,t) = \frac{1}{8\pi^2} \cdot \sum_{\alpha,\gamma=1}^{2} \rho_\alpha \left\{ \iint d\Omega_0 d\Omega \alpha_{zz}^\gamma(\Omega,t)\alpha_{zz}^\alpha(\Omega_0,0) \right.$$

$$\left. \widetilde{P}_{\gamma\alpha}(K,\Omega,\Omega_0,t) \right\} \tag{41}$$

The $\widetilde{P}_{\gamma\alpha}$ may be expanded in terms of rotation matrices with coefficients dependent only on K and t,

$$\widetilde{P}_{\gamma\alpha}(K,\Omega,\Omega_0,t) = \sum_{J,K,M} B_{\gamma\alpha}^J(K,t)D_{K,0}^J(\Omega)D_{K,0}^{J*}(\Omega_0) \tag{42}$$

Solving Eqs. (38) by Fourier space and Laplace time transformation, and using Eqs. (11) and (42), we find

$$C_{zz}(K,\omega) = \frac{1}{\pi} \sum_{\gamma,\alpha=1}^{2} \rho_\alpha \alpha_I^\gamma \alpha_I^\alpha Re\tilde{B}^0_{\gamma\alpha}(K,s=i\omega)+$$

$$\frac{2}{15\pi} \sum_{\gamma,\alpha=1}^{2} \rho_\alpha \alpha_0^\gamma \alpha_0^\alpha Re\tilde{B}^2_{\gamma\alpha}(K,s=i\omega) \qquad (43)$$

where $Re\tilde{B}^J_{\gamma\alpha}(K,s=i\omega)$ is the real part of the Laplace transform of $B^J_{\gamma\alpha}(K,t)$ with s set equal to $i\omega$. The $\tilde{B}^0_{\gamma\alpha}(K,s)$ and $\tilde{B}^2_{\gamma\alpha}(K,s)$ are given by

$$\tilde{B}^J_{11}(K,s) = \frac{(s + K^2 D_2 + c_{J2} + k_{12})}{\Delta_J(s)}$$

$$\tilde{B}^J_{21}(K,s) = \frac{k_{21}}{\Delta_J(s)}$$

$$\tilde{B}^J_{12}(K,s) = \frac{k_{12}}{\Delta_J(s)}$$

$$\tilde{B}^J_{22}(K,s) = \frac{(s + K^2 D_1 + c_{J1} + k_{21})}{\Delta_J(s)} \qquad (44)$$

where

$$\Delta_J(s) = (s + K^2 D_1 + c_{J1} + k_{21}) \boldsymbol{x} (s + K^2 D_2 + c_{J2} + k_{12})-$$

$$k_{21}k_{12}$$

and

$$c_{J\gamma} = \begin{cases} 0 & \text{for } J = 0 \\ 6\Theta_\gamma & \text{for } J = 2 \end{cases}$$

A similar calculation for $C_{yz}(K,\omega)$ gives

$$C_{yz}(K,\omega) = \frac{1}{10\pi} \sum_{\gamma,\alpha=1}^{2} \rho_\alpha \alpha_0^\gamma \alpha_0^\alpha Re\tilde{B}^2_{\gamma\alpha}(K,s=i\omega) \qquad (45)$$

Previous authors have neglected the second term on the right hand side of Eq. (43). This amounts to assuming that the molecules are optically isotropic. If this is not the case and only the isotropic term is wanted, it is clear that both C_{zz} and C_{yz} must be measured and C^I_{zz} extracted by use of the relation

$$C^I_{zz}(K,\omega) = C_{zz}(K,\omega) - \frac{4}{3}C_{yz}(K,\omega) \qquad (46)$$

Discussions of Eqs. (43) and (45) and explicit formulae for various limiting cases have been given by several authors[14,16]. We note here that the results reduce to the correct two component form when the reaction is turned off, i.e., $k_{12} = k_{21} = 0$. When

the two species have equal polarizabilities and diffusion coefficients, the results reduce to those of a single species scatter of concentration $\rho = \rho_1 + \rho_2$ with no rate constant contributions to the line width[12]. It is clear on the basis of physical
arguments that this must be true since, in such a case, the light
wave has no way of "seeing" the chemical reaction.

The results are especially simple when the reaction rates are
much faster than the translational and rotational diffusion rates.
For instance, if we neglect the contributions of rotation and translation to the line widths, Eqs. (43) and (45) become

$$C_{zz}(\kappa,\omega) = \frac{1}{\pi}(B + \frac{2D}{15})\frac{\lambda}{\omega^2 + \lambda^2} \tag{47}$$

and

$$C_{yz}(\kappa,\omega) = \frac{1}{10\pi}D\frac{\lambda}{\omega^2 + \lambda^2}$$

where $\lambda = k_{12} + k_{21}$ and

$$B = \frac{\rho_1 \rho_2}{\rho}(\alpha_I^{(1)} - \alpha_I^{(2)})^2$$

$$D = \frac{\rho_1 \rho_2}{\rho}(\alpha_0^{(1)} - \alpha_0^{(2)})^2$$

V. Polydisperse Solutions

The results given above must be modified when dealing with
polydisperse polymer solutions. In this case the solution may be
considered to be a multicomponent system. Each component will have
different dynamical constants and structure, giving rise to its
own spectrum as described in Section III. Hence, to obtain the
resultant spectrum, the equations of Section III must be averaged
over the whole polymer distribution. Quantities such as the
molecular weight (or size) dependence of the rotational and translational diffusion coefficients and intramolecular relaxation
times as well as the distribution function must be known. One
might also, under favorable circumstances, expect to use the light
scattering spectrum to determine the molecular weight distribution
of the sample. The scattering spectrum might, in addition, be used
to study aggregation in biological materials.

A calculation of the effects of polydispersity on light
scattering from polydisperse rods and gaussian coils has been
performed using the Schulz two parameter unimodal molecular weight
distribution function[18].

VI. Discussion

The wide variety of dynamic information obtainable from light scattering spectroscopy is providing the stimulus for much experimental and theoretical work in this field. So far only a relatively small amount of experimental work has been published and much of this work has been aimed at demonstrating that the technique is practical rather than obtaining new and useful information. Only one experiment has been performed to date on the light scattering spectra of chemically reacting systems[19]. None has yet been performed on macromolecular reactions. Theoretically only the simplest models have been studied. For instance, the detailed macromolecular theory presented above could be used to obtain more than purely formal results only for dilute solutions. Applications to more concentrated solutions should be made. The problem of the calculation of intramolecular form factors for the parts of the spectra dependent on the anisotropic spherical polarizability components should be considered. All effects from these form factors were ignored in the theories presented in Sections III and IV. Furthermore, some molecular theories of polarizability changes accompanying macromolecular chemical reactions and conformation changes should prove to be very useful.

In this review, we have not discussed instrumentation or experimental technique. Two excellent reviews of these subjects are forthcoming[20,21]. We should mention that a major advantage of light scattering over other techniques for obtaining dynamic information about polymers is that one studies the time decay of the natural fluctuations of a system that is macroscopically in equilibrium. Thus, no external force of any kind is needed to perturb the system from equilibrium. This should be especially important in studying delicate biological materials.

There are many possible applications of light scattering spectroscopy to nonequilibrium biological problems. For instance, light scattering might prove itself to be very useful in the study of membrane transport phenomena as well as in the study of natural, microscopic, oscillatory processes in living systems. To the author's knowledge, no applications of this type have yet been made.

REFERENCES

1. See, for instance, the articles reprinted in Light Scattering from Dilute Polymer Solutions, D. McIntyre and F. Gornick, Eds. (Gordon and Breach Science Publications, Inc., New York, 1964).
2. R. Pecora, Ph.D. Thesis, Columbia University, 1962. See also, R. Pecora, J. Chem. Phys. 40, 1604 (1964).
3. H. Z. Cummins, N. Knable and Y. Yeh, Phys. Rev., Letters, 12 150 (1964).

4. S. B. Dubin, J. H. Lunacek and G. B. Benedek, Proc. Natl. Acad. Sci. (U.S.) 57, 1164 (1967).

5. F. T. Arecchi, M. Giglio and U. Tartari, Phys. Rev. 163, 186 (1967).

6. H. Z. Cummins, F. D. Carlson, T. J. Herbert and G. Woods, Biophys. J. 8, A95 (1968). Abstract TD1.

7. A. Wada, N. Suda, T. Tsuda and K. Soda, J. Chem. Phys., 50, 31 (1969).

8. N. C. Ford, W. Lee and F. E. Karasz, J. Chem. Phys., 50, 3098 (1969).

9. M. J. French, J. C. Angus and A. G. Walton, Science 163, 345 (1969).

10. R. Pecora, J. Chem. Phys., 43, 1562 (1965); 49, 1032 (1968).

11. R. Pecora, J. Chem. Phys. 48, 4126 (1968).

12. R. Pecora, J. Chem. Phys. 49, 1036 (1968).

13. R. Pecora, Macromolecules, 2, 31 (1969).

14. B. J. Berne, J. M. Deutch, J. T. Hynes and H. L. Frisch, J. Chem. Phys., 49, 2864 (1968).

15. L. Blum and Z. M. Salzburg, J. Chem. Phys., 48, 2292 (1968); 50, 1654 (1969).

16. B. J. Berne and R. Pecora, J. Chem. Phys. 50, 783 (1969); 51, 475 (1969).

17. K. A. Valiev and L. D. Eskin, Opt. Spectrosc., 12, 429 (1962); K. A. Valiev, ibid., 13, 282 (1962).

18. Y. Tagami and R. Pecora, J. Chem. Phys., 51, 3293, 3298 (1969).

19. Y. Yeh and R. N. Keeler, J. Chem. Phys., 51, 1120 (1969).

20. Y. Yeh and R. N. Keeler, Quart. Rev. of Biophys. In press.

21. H. Z. Cummins and H. L. Swinney, in Prog. in Optics, Vol. 8, Ed. E. Wolf (North Holland Publishing Co., Amsterdam). In press.

HYPOCHROMISM IN DIMERS

Mitchel Weissbluth

Department of Applied Physics

Stanford University, Stanford, California 94305

The transition of a nucleic acid from a random coil to a helical configuration is accompanied by a reduction in total integrated
intensity in the 2600 A° absorption band. The general shape of the
spectrum remains unaltered. This effect is known as hypochromism
— it has also been observed in the absorption bands of nucleic
acid polymers and polypeptides.

The current picture of the mechanism of hypochromism contains
three elements: (1) Each molecule of a polymer is subjected to a
local field which is a superposition of fields from two types of
sources — one is the incident light beam and the other is the
induced dipole moments of neighboring molecules. Since the local
field with which each molecule interacts may be quite different
from the optical field, alterations in the absorption spectrum of
the polymer may be expected. As long as the molecules have fixed
orientations relative to one another the local fields achieve a
steady state. If, on the other hand, the relative orientations
fluctuate in a random manner, the effect of one molecule upon its
neighbors — insofar as the absorption spectrum is concerned —
averages to zero. (2) Although there may exist several interactions
which contribute to spectral changes upon ordered aggregation, it
is thought that for the biological molecules the dominant effect of
one molecule upon its neighbors is through the Coulombic interaction
between transition dipoles. The latter are defined as matrix elements of the dipole moment operator between two distinct molecular
states. Moreoever, it is usually sufficient to consider only those
transitions which excite a molecule from its ground state. (3)
The interaction is of such a nature that when one absorption band
loses intensity (hypochromism) there is a compensating increase in

absorption intensity in other bands (hyperchromism). It is as if
one band borrows intensity from other bands while the total absorption intensity, over all bands, remains constant. An equivalent
statement is that the sum rule for oscillator strengths taken over
all bands is satisfied.

A brief review of the main theoretical lines is presented
within the framework of the very simplest model which is capable
of exhibiting hypochromism. Such a model is a dimer consisting
of two identical molecules each having a ground state and two non-
degenerate excited states.

(1) Classical [1] : A set of coupled equations are written
which express the induced dipole moment of each molecule in its
own local field which, of course, includes the effect from the
induced dipole moment of the neighbor. These are

$$\mu_a = \alpha(\omega)\left[\vec{\ell}_a \cdot \vec{E} - V_{ab}\mu_b\right]$$

$$\mu_b = \alpha(\omega)\left[\vec{\ell}_b \cdot \vec{E} - V_{ab}\mu_a\right] \tag{1}$$

in which $\vec{\mu}_i = \mu_i \vec{\ell}_i$ is the induced dipole moment in the i-th
monomer $(i = a, b)$; $\alpha(\omega)$ the polarizability of each monomer;
$\vec{E}$ the electric field of the incident light wave: $\mu_a\mu_b V_{ab}$ the
Coulomb interaction between monomers. Eqs. (1) yield an expression for the polarizability of the dimer,

$$\alpha_d = \frac{2\alpha(\omega)}{1 + \alpha(\omega) V_{ab}} \tag{2}$$

whose imaginary component is responsible for absorption and is
related to the extinction coefficient or oscillator strength. If
each molecule has only one resonant frequency — the classical
analogue of a molecule with a ground state and a single excited
state — it is found that the oscillator strength of the dimer
relative to the oscillator strengths of the two monomers remains
unchanged, although there is a frequency shift. However, when
the polarizability of each monomer is permitted to have two resonant frequencies,

$$\alpha(\omega) = \frac{e^2}{m} f \left[\frac{1}{\omega_1^2 - \omega^2} + \frac{1}{\omega_2^2 - \omega^2} \right] \qquad (3)$$

corresponding to the existence of two excited states, the oscillator strength of the dimer becomes

$$fd = \frac{2m^2}{e^4 V_{eb} f} \left[\frac{1}{\left(\omega_1^2 - \omega_d^2 \right)^2} + \frac{1}{\left(\omega_2^2 - \omega_d^2 \right)^2} \right]^{-1} \qquad (4)$$

Eq (4) exhibits hypochromism (and hyperchromism) ; moreover, it becomes quite clear from an analysis of the behavior of the singularities of the dimer polarizability function [2] how it is possible for a band to show hypochromism while retaining its shape with little or no shift in its peak position.

(2) Exciton [3,4] : This was historically the first approach to explain hypochromism. In this formalism, which uses time-independent perturbation theory, the stationary states of the dimer are computed and the oscillator strengths are obtained from matrix elements for transitions between dimer states. In zero order (non-interacting molecules) the dimer ground state is non-degenerate while excited states are doubly-degenerate. Turning on the perturbation (dipole-dipole interaction between transition dipoles) has essentially no effect on the ground state but each initially-degenerate excited state separates into two states. One finds, in this case that although there may be spectral shifts, there is no hypochromism. For the latter to appear it is necessary to go to first order wave functions and to compute matrix elements for transitions among them. It is then seen that hypochromism can occur provided each molecule has at least two excited states; the sum rule is obeyed. It is necessary to remark that although excitons appear in the formalism the phenomenon of hypochromism does not depend on specific properties of excitons but rather on the interaction between absorption bands as expressed through the Coulombic interaction between transition dipoles.

(3) Field Theory [5,6]: This approach is based on the response function formulation of time-dependent perturbation theory [7,8]. For the case being considered, the dimer polarizability is expressed by

$$\alpha_d = \int_0^\infty k(\tau)e^{i\omega\tau}d\tau \qquad (5)$$

where $k(\tau)$, the response function, is to be calculated as a
perturbation expansion. The calculation is facilitated by writing
the Hamiltonian of the system in terms of creation and annihilation
operators which are defined in such a manner that when they act on
a wave function in an occupation number space they create or anni-
hilate an excitation in a particular molecule. The operators
obey a set of commutation rules which are, rigorously, neither
of the Bose nor Fermi type; nevertheless an approximation to Bose
operators may be made and the Hamiltonian, which includes the
interaction between molecules, may then be written in a convenient form. The time-dependent perturbation contains two terms —
one is the Coulomb interaction between the two monomers arising
from induced dipole moments and a second term which describes the
influence of the external time-varying electric field of the light
wave. In the interaction representation, the perturbation satis-
fies a time-dependent Schrodinger equation whose solution may be
expressed in integral form with a kernel or Green's function bearing a close relationship to the polarizability.

Although these methods are quite different in their formal
aspects they are, nevertheless, remarkably consistent with one
another.

This work was supported by ONR contract Nonr-225(87) and NIH grant
GM16690.

References

1. H. De Voe, J. Chem. Phys. 41, 393 (1964)
2. A. Herzenberg and A. Modinos, Proc. Phys. Soc. 87, 597 (1966)
3. I. Tinoco, Jr., J. Chem. Phys. 33, 1332 (1960); 34, 1067 (1961)
4. W. Rhodes, J. Am. Chem. Soc. 83, 3609 (1961)
5. R. Hoffman, Rad. Res. 20, 140 (1963)
6. W. Rhodes and M. Chase, Rev. Mod. Phys. 39, 348 (1967)
7. R. Kubo and K. Tomita, J. Phys. Soc. Japan 9, 888 (1954)
8. R. Kubo, J. Phys. Soc. Japan 12, 570 (1957)

QUANTUM MECHANICALLY BASED RULES FOR THERMAL AND PHOTOCHEMICAL REACTIONS

E. M. Evleth

Division of Natural Sciences, University of California,

Santa Cruz, California 95060

INTRODUCTION

There are at least three approaches necessary for the application of quantum mechanical considerations to the analysis of the mechanisms of thermal and photochemical reactions. These approaches are: i) generalized rules for correlating the electronic states of the starting materials and the products; ii) direct computational estimates of the potential energy surfaces and other properties of the systems under consideration, and iii) qualitative rules by which the possible shapes of the potential energy surfaces can be predicted or the possible course of the reactions can be rationalized. Approaches i and iii provide a theoretical base for building the necessary qualitative understanding of thermal and photochemical reactions and ii provides a route for partially testing that understanding. Since there is no comprehensive discussion of all these factors in the literature this paper will attempt to correlate these approaches with what has been found to date. No attempt is made, however, to give a complete review of the literature on this subject. Particular emphasis has been given to the recent literature.

GENERALIZED RULES FOR CORRELATING ELECTRONIC STATES

The rules for correlating the symmetries of the electronic states involved in thermal and photochemical decomposition of diatomic triatomic, and a limited number of polyatomic molecules have been discussed in detail by Herzberg (1,2). These rules are based both on symmetry considerations and the law of conservation of spin angular momentum. The usefulness of these rules is

limited to systems of high symmetry. What is interesting is that
for many small molecules (e.g., CO_2, N_2O (3,4)) the potential
energy surface of the ground state of the filled shell molecule is
not adiabatically connected to the ground states of fragmented
species. However, it seems extremely likely that in a large molecular system, such as toluene, the ground state is adiabatically
connected to the ground state of benzyl radical and hydrogen atom.
It also seems likely that the relative energies of toluene and
benzyl radical in the ground and lower excited states can be adequately approximated using as the basis set the valence orbitals
of carbon and hydrogen. In the higher electronic states (i.e.,
Rydberg (5)) the incorporation of the excited states of carbon
and hydrogen into the basis set for a molecular computation would
be necessary. It also seems likely that all first row elements
could be incorporated into molecular computations using only the
valence orbitals. The conclusion is that the thermal and photochemistry of small and large polyatomic molecules are differentiated. From these considerations it is possible to formulate the
rule that for large low symmetry filled shell systems the ground
states of the starting materials and products lie on the same potential energy surface. Such a rule has already been formulated
for isomerization reactions (6). However, severe symmetry re-
strictions must accompany this rule. In situations where a filled
shell molecule thermally decomposes into two or more filled shell
molecules symmetry arguments dictate that the rule is absolute.
In cases where the molecular fragments are not filled shell species
(neutral or ionic) the rule is not absolute and important excep-
tions may exist.

Using these above principles it is possible to formulate
some generalized rules correlating the electronic states of the
original and final systems for thermal and photochemical breakup
of large molecular systems. The hypothetical example of a large
polyatomic filled shell molecule AB is chosen. It shall be chosen
to be a neutral species because the rules for AB being a filled
shell anion or cation are the same. Five different possibilities
are discussed: A) homolytic fragmentation yielding radicals; B)
homolytic fragmentation yielding singlet and triplet species; C)
heterolytic fragmentation yielding singlet and triplet ions; D)
heterolytic fragmentation yielding radical anions and radical cations; and E) isomerization. The reactions under consideration
can either be considered going from right to left or left to
right. It should be kept in mind that these rules are based on
purely adiabatic considerations and are hypothesized for the condensed phase only.

A) Homolytic fragmentation yielding radicals.

    i) Excited singlet AB yields an excited radical

$$^1AB^* \quad \rightarrow \quad ^2A + ^2B^*$$

ii) Triplet AB Yields Directly or Indirectly Ground State Radicals

$$^3AB^* \quad \rightarrow \quad ^2A + ^2B$$

iii) Ground State AB Yields Ground State Radicals

$$^1AB \quad \rightarrow \quad ^2A + ^2B$$

The above rule is based on the diatomic model where AB is decomposing by the breaking of one bond. The combination of doublet species may yield triplet or singlet species. This may occur from either the ground state or excited states of the doublet species. Considerations dictate that along certain symmetry coordinates the lowest spectroscopic triplet may not be adiabatically connected to the ground state of the doublet species but to the excited state. However, in polyatomic cases curve crossing does not exist as such and the lowest spectroscopic triplet will be connected with the radical products. If the spectroscopic triplet lies at higher energies than the bond dissociation energy for thermal AB bond breaking then radical formation can occur. However, unlike the H_2 dissociation case the potential energy surface for triplet decomposition does not have to be totally repulsive. It can be, however, and triplet decomposition can occur with or without an activation energy. In the former case phosphorescence and monitoring the triplet concentration (by triplet-triplet absorption) is conceivable besides directly determining the activation energy of triplet decomposition. Because of the problem of the short lifetime of triplet species the activation energy must be in the order of 10 kcal or less if decomposition is to occur at room temperature in the condensed phase.

An additional consequence of the above rule is that the thermally generated radicals and triplet photochemically generated radicals are the same but the excited singlet generated radicals are different (one of them being electronically excited). However, luminescence from excited radicals is only observed in far ultraviolet photochemistry of small polyatomic molecules (1,2,7). Since bond dissociation energies are in the 50-100 kcal region and it is probable that the excited radical species are in the same energy region, the generation of excited radicals from a typical molecule AB would have to be endothermic by at least 100 kcal. Thus emission from excited radical species generated directly in a photochemical reaction would have to occur from singlets excited at energies higher than 250 mμ. It is likely that decomposition in the excited singlet state would require activation energies in excess of those attainable thermally in the condensed phase at room temperature. It seems more likely that decomposition through the singlet route would require the intervention of the non-adiabatic process (8). It is generally, although not always observed, that photochemically generated radicals result from decomposition through a triplet mechanism (9-18).

   B) Homolytic fragmentation yielding singlet and triplet fragments.

   i)   Excited Singlet AB yields an Excited Singlet

$$^1AB^* \rightarrow \ ^1A \ + \ ^1B^*$$

   ii) Triplet AB yields a Triplet Fragment

$$^3AB^* \rightarrow \ ^1A \ + \ ^3B^*$$

   iii) Ground State AB yields Ground State Fragments

$$^1AB \rightarrow \ ^1A \ + \ ^1B$$

   The decomposition of AB in the mode described above can be thought of occurring through the cleavage of two bonds joining fragments A and B. The resulting fragments can either be ground state filled shell or in their excited states. Regardless of the route of decomposition the filled shell axiom plus configuration interaction states that the ground state of all fragments must be totally symmetric and thus lie on the same potential energy surface.

   In the excited state this type of fragmentation is manifested experimentally in several different forms. Excimer formation, mixed excimer formation, and charge transfer complexes are conceptually related (20,21). Simple excimer formation, A_2, does not occur in the ground state, probably not in the triplet, but does in the excited singlet (20). In charge transfer complex formation, AB, all three states may be stable (21). The logical limit of excimer, mixed excimer, and charge transfer complex formation is where two distinct bonds are formed (at closer geometries than occur with complex formation) to give photodimers or photoadducts. Generally the excited singlet and triplet states of photodimers and photoadducts are higher above the ground state of AB than of A and B separated at infinite distance. So unless the formation of AB is highly exothermic both processes i and ii are unlikely to occur directly (in reverse) for either photodimerization or photocycloaddition. A possible compromise for triplet mechanism is a two-step process in which an intermediate biradical, $\cdot$A-B$\cdot$, is formed as an intermediate. Under such circumstances the reaction becomes analogous to that discussed in Rule A where the potential energy surface of the triplet two-step process, $^1A + ^3B \rightarrow \ \cdot$A-B$\cdot$, comes into close proximity with the ground state surface. Such a two-step excited singlet mechanism is unlikely since it would require the formation of an intermediate excited biradical $\cdot$A-B$\cdot$*. In both the above cases non-radiative processes must determine the formation of photodimers or adducts. Generally this conclusion is supported by detailed observations (22-25).

    C) <u>Heterolytic fragmentation to give singlet and triplet states</u>.

        i) Singlet Excited AB yields an Excited Singlet

$$^1AB^* \rightarrow \quad ^1A^+ + \, ^1B^{-*}$$

        ii) Triplet AB yields a Triplet Fragment

$$^3AB^* \rightarrow \quad ^1A^+ + \, ^3B^{-*}$$

        iii) Ground State AB yields Ground State Fragments

$$AB \rightarrow \, ^1A^+ + \, ^1B^-$$

The above three sequences of ground state, triplet, and excited singlet ionization has been observed and confirmed in the thermal and photoionization of organic acid and bases (26). Examples where A^+ is not a proton have not been confirmed for processes i and ii although ground state ionization or solvolysis is a common event. Monitoring fluorescence, phosphoresence, or triplet-triplet absorption would confirm the presence of processes i and ii in a particular photochemical reaction.

An important point to stress is that Rules A and C govern the breaking of a single bond in a homolytic or heterolytic fashion. In the gas phase heterolytic fragmentation does not occur in the ground state and for it to occur in the condensed phase results from the inclusion of solvent. In the gas phase the annihilation of two filled shell ions will result in the generation of an excited singlet. Thus the solvent plays the role of making a reaction which would normally generate excited species yield a solvated ground state specie. Thus the solvent can reorder the respective roles of Rules A and C in a particular photochemical reaction. This has been shown for the case of the photolysis of 3- and 4-methoxybenzyl acetates in various solvents (27). What is not clear from a simple analysis is whether both homolytic and heterolytic fragmentation can occur simultaneously in a particular solvent from the same spectroscopic state.

    D) <u>Heterolytic fragmentation to yield radical ions</u>. This type of fragmentation defies a simplistic analysis. At first glance this type of fragmentation looks like a combination for Rules A, and C. Thus, one can write the following equations:

$$^1AB^* \quad \rightarrow \quad ^2A^+ \quad + \quad ^2B^{-*}$$

$$^3AB^* \quad \rightarrow \quad ^2A^+ \quad + \quad ^2B^-$$

$$^1AB \quad \rightarrow \quad ^2A^+ \quad + \quad ^2B^-$$

The reaction is largely observed from right to left in the annihilation of radical anions and cations of the same species (electroluminescence). Although cation-anion annihilation is observed to give directly triplet species the direct formation of excited singlet species is less clear (28,29). On the other hand if solvent conditions are proper a strongly oxidizing specie, B could react with an electron donar A to give a charge transfer complex, AB, which could in turn break up into ground state radical ions $^2A+$ and $^2B-$. Thus, as with the competition of Rules A and C, solvent plays an important conceptual part. Conceptually (not actually) the annihilation of A^+ and A^- (such as with anthracene) can occur in a two-step process. First the annihilation of the radical portion by forming a two electron bond

$$^2A^+ \quad + \quad ^2A^- \rightarrow \quad ^{1,3}(^+A\text{---}A^-)$$

which is analogous to Rule A giving a zwitterion in the ground or triplet state (in the statistical ratio of 1 to 3, respectively). Now the annihilation of the ionic portion would go according to Rule C if the solvent is strongly solvating to give the collision complex A_2 in the ground or triplet state. However, under weak solvent conditions annihilation could give excited singlet (or triplet) A_2^*, this conceptually is analogous to the annihilation of H^+ and H^-. If the annihilation of ions $^2A^+$ and $^2A^-$ occurs along certain symmetrical coordinates, it is possible to impose some symmetry arguments on the correlation of the states of the ions with those of A_2 and, finally, the actually observed excited states of A resulting from the breakup of A_2. However, there is no assurance that the molecules actually react along these coordinates.

E) Isomerization. Isomerization reactions, $M \rightarrow N$, can proceed in several different fashions which are the intermolecular counterparts of Rules A-D. Conceptually the reaction can proceed in a concerted fashion with no minimum on the potential energy surface region of consequence except at the product points M and N. In addition a multistage process is conceivable where one or more minima are present. Both calculations (31-33, 42) and experimental data (34-41) indicate that the thermal isomerization of ethylene and substituted ethylenes is a concerted process but that photoisomerization in the triplet and possibly singlet could be a one or two-step process. In the dienes (31) this can conceptually be written occurring through the triplet state in the following manner:

$$^3M^* \quad \rightarrow \quad \cdot A\text{-}B\cdot \quad \overset{a}{\nearrow} \, ^3N^*$$
$$\underset{b}{\searrow} \, N$$

In dienes generation of the biradical from ^{3}M* is exothermic. Generation of ^{3}N* from the biradical (process a) is endothermic thus intersystem crossing (process b) is preferred as the actual route to isomerization. This process is the intramolecular counterpart of those discussed in Rule A since the potential energy surfaces of the ground and triplet states of M come into close proximity at the geometry of the biradical (31).

The ionic counterpart of cis-trans isomerization (Rule C) has also been described (31). Here a zwitterion is the conceptual intermediate.

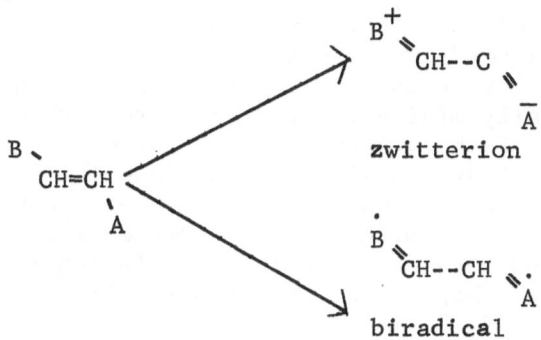

Twisting about the central single bond in butadiene occurs in the ground state with a low thermal barrier but a high barrier in the triplet state (39). Actual calculations (31) indicate that the behavior of this system is the intramolecular counterpart of that discussed in Rule B. When the geometrical distortion of the molecule breaks it up into filled shell connected fragments (^{1}M $\rightarrow$ ^{1}A- ^{1}B) the expectation is for a low ground state but higher excited state thermal barriers.

In isomerization of the valence type the expectation is that the mechanisms can occur through intramolecular counterparts of those discussed in Rules A through D. In those systems in which the pi-electron network is partially or completely converted into a sigma bond network (sometimes highly strained) it is energetically impossible (40-42) for excited M* to generate N*, thus internal conversion or intersystem crossing must determine the partitioning of product formation N and return to ground state M (6,40,41). This is true for the butadiene-cyclobutene-bicyclobutane introconversion (40,42). Emission from N* is not observable in such systems. It should be observable if such processes are exothermic or nearly isoenergetic. Generally such emission is only observed in those cases where hydrogen transfer from one location to another in the molecule occurs (43,44).

COMPUTATIONAL APPROACHES TO ESTIMATING POTENTIAL ENERGY SURFACES

The two major methods used for the approximate solution of the Schrödinger Equation are the valence bond and molecular orbital approaches. Although neither method is fundamentally superior to the other historically the molecular orbital method has been computationally simpler. Until recently most calculations have not been concerned with estimating molecular wavefunctions at other than equilibrium geometries. Classically the simple molecular orbital approach predicts that the potential energy surface for the ground state of H_2 is adiabatically connected with H^+ and H^- (ground state) at the dissociation limit when, in fact, ground state hydrogen atoms should be the products. This inadequacy in the single determinant Hartree-Fock self-consistent field molecular orbital wavefunction is partially cured by the inclusion of a larger basis set and configurationally mixing the ground state with doubly and higher excited states of the molecular wavefunction (45,46). Other techniques (47-50) are available. The imposition of configuration interaction on the ground state molecular orbital wavefunction is a necessity for calculation attempting to describe the potential energy surface at the bond dissociation limit. Obviously the role of such higher configurations will be important for the excited states of the molecule (51-56).

In recent years an increasing number of potential energy surface calculations have appeared in the literature (57-68). To a large degree configuration interaction considerations have been neglected. Obviously in certain situations such as low barrier isomerization reactions configuration interaction would be of absolute but not relative importance. The expectation is that where a high thermal barrier exists for a ground state or excited state reaction configuration interaction would be of great relative importance. Below we shall discuss a few of the most widely used semiempirical and nonempirical techniques with the view of placing limits on their ability of yielding theoretically reliable potential energy surfaces.

Extended Huckel Method

This method (57) specifically neglects electron repulsion terms. This semiempirical molecular orbital method does include all valence electrons. Therefore low barrier isomerizations and rotational barriers can be adequately treated under proper parameterization. The multiplicity of the excited states of filled shell systems can not be energetically differentiated because of neglect of electron-electron repulsion integrals. The technique is, therefore, not applicable to calculating surfaces in the excited state. It is possible that configuration interaction is not of relative importance in calculated ground state potential energy surfaces for the breakup of the molecule AB into filled shelled fragments A and B. Likewise the addition of ion or a radical to a filled shell configuration is subject to treatment (64,65). Thus the extended Huckel method is potentially capable of handling some ground state processes discussed in Rules B, C, and F.

CNDO and Other Methods

The Pople-Santry-Segal/CNDO and related methods (69-71) are semiempirical self-consistent field molecular orbital techniques which include electron repulsion terms. To date configuration interaction has generally been applied only at the level of estimating excitation energies (70,72,73). The imposition of doubly excited and higher configurations has generally been neglected (60,61) in potential energy surface calculations. However the neglect of configuration interaction would render such calculations meaningless for ground state potential energy surfaces which yield radicals or radical ions and for certain high barrier isomerization reactions. Calculations on potential energy surfaces for excited states need not require inclusion of configuration interaction if handled using open shell methods.

Several attempts have been made to estimate the energy of certain configurations using a Pariser-Parr-Pople method applied to non-planar systems (31-33). Here, also, configuration interaction is important (31) for high barrier calculations. Such effects are important (31) even though neglected in some calculations (32,33,74).

Nonempirical Methods

A number of nonempirical calculations of potential energy surfaces of simple thermal reactions have recently appeared (49,58,59,66,67,68,75). Largely these have been low barrier calculations and configuration interaction has been neglected. In ethylene (58,59) correct calculation of the ground and excited state surfaces absolutely require the inclusion of configuration interaction.

QUALITATIVE RULES BASED ON SYMMETRY AND OTHER CONSIDERATIONS

As previously mentioned Herzberg (1,2) has discussed in detail the correlation between the symmetry of the starting material and products for thermal and photochemical processes. Such rules are based on the United Atom treatment. In addition Walsh (76) has described a qualitative method for predicting the shape of triatomic molecules in the ground and excited states. These rules are based on the conservation of orbital symmetry. More recently

Longuet-Higgins and Abrahamson (77) analyzed the observations of
Woodward and Hoffmann (78) with regard to the stereochemical differences between reactions taking place in the ground and excited
states. Longuet-Higgins and Abrahamson (77) based their arguments
on the conservation of orbital symmetry. Woodward and Hoffmann (78-
81) expanded this interpretation beyond the area of pure symmetry
and have hypothesized a series of general rules based on limited
orbital analysis which enable a qualitative prediction as to the relative barriers in the ground and excited states for different
stereochemical courses of a particular reaction. The Woodward-Hoffman Rules have now been well reviewed (79-81).

One of the simplest applications of orbital symmetry rules
demonstrating the importance of configuration interaction occurs in
the cis-trans isomerization of dideuteroethylene. The isomerization
is known to occur in either thermal or photochemical processes (34,
82).

$$
\underset{C_{2h}}{\overset{\displaystyle D\diagdown_{}\quad \diagup H}{\underset{H\diagup^{}\quad\diagdown D}{C=C}}}
\quad\longrightarrow\quad
\underset{\underset{C_2}{intermediate}}{\overset{\displaystyle D\diagdown_{}\quad \diagup H}{H\!\!\rightarrow\!\!C-C\diagdown D}}
\quad\longrightarrow\quad
\underset{C_{2v}}{\overset{\displaystyle D\diagdown_{}\quad \diagup D}{\underset{H\diagup^{}\quad\diagdown H}{C=C}}}
$$

The symmetry species of cis- and trans-dideuteroethylene are
C_{2v} and C_{2h}, respectively. Twisting about the essential double bond
produces an intermediate of C_2 symmetry along the reaction coordinate.
Of all the possible symmetry operations only the C_2 operation (rotation by 180° about the 2-fold axis) is common to all three molecules.
It follows that the molecular orbitals for these three molecules will
be either symmetric (s) or antisymmetric (a) to this symmetry operation. The consequences of these operations are shown in Figure 1.
Focusing just on the 2p orbitals of carbons 1 and 2 which comprise
the original pi-orbitals of the planar molecule, the effect of this
twisting on the one electron orbitals is shown in the orbital diagram.
These individual one electron orbitals of cis-and trans-D_2-ethylene
have the following symmetries with respect to the C_2 operation:
trans, pi(s), pi*(a); cis, pi (a), pi*(s). Thus the pi-bonding orbitals of cis-and trans-D_2-ethylene are not symmetry connected; the
pi-bonding orbital of one specie adiabatically changes into the pi-
bonding orbital of the isomer. The state diagram of Figure 1 shows
that the ground and doubly excited states of both the cis and trans
ethylenes have the same total symmetry (sxs = S, and axa = S), and
thus the potential energy curves can not cross. In the region of
90° twist the ground the the doubly excited states of the C_2 configuration of ethylene strongly interact (configuration interaction)
and the curves fail to cross. Although direct computation (58,59)
shows this to be the case symmetry considerations alone show that
doubly excited configurations can not be neglected in the calculation of ethylene. However such considerations can not say anything

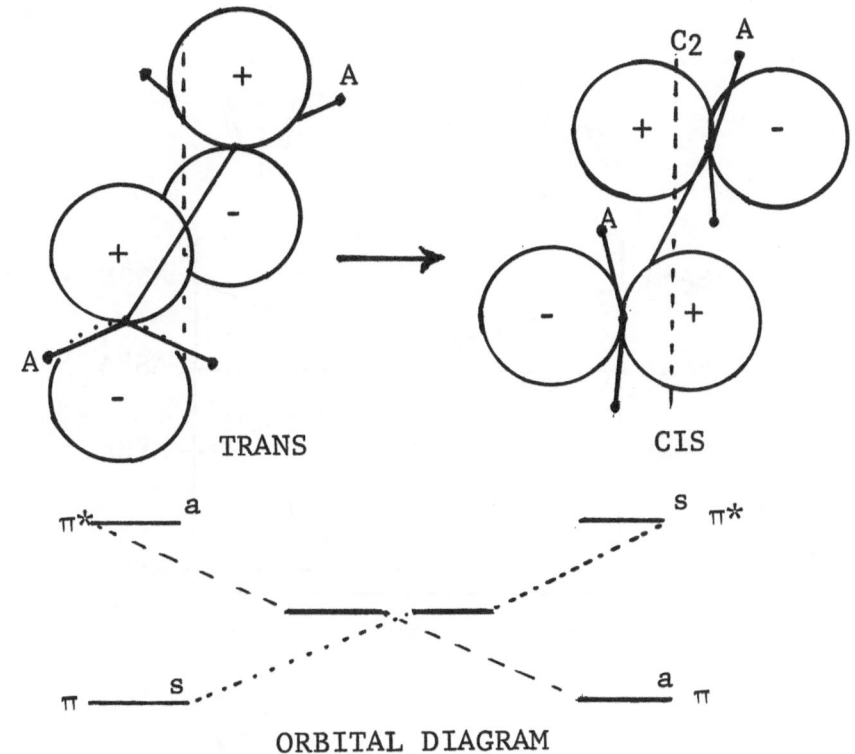

ORBITAL DIAGRAM

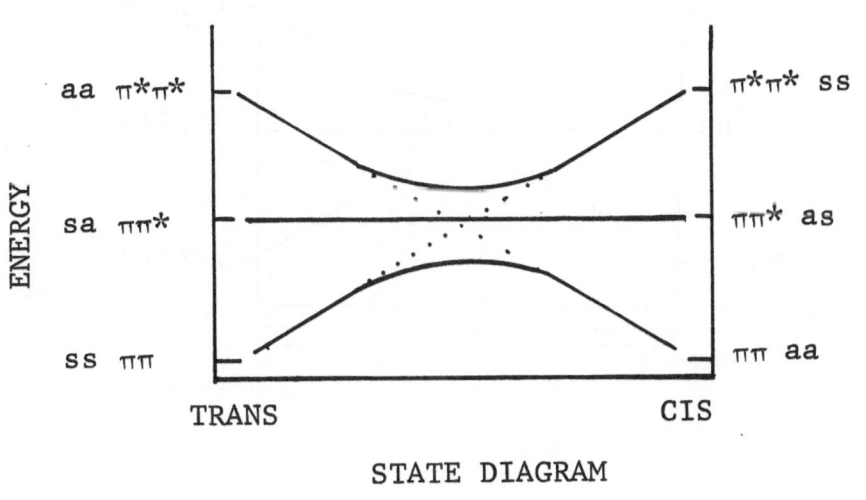

STATE DIAGRAM

Figure 1

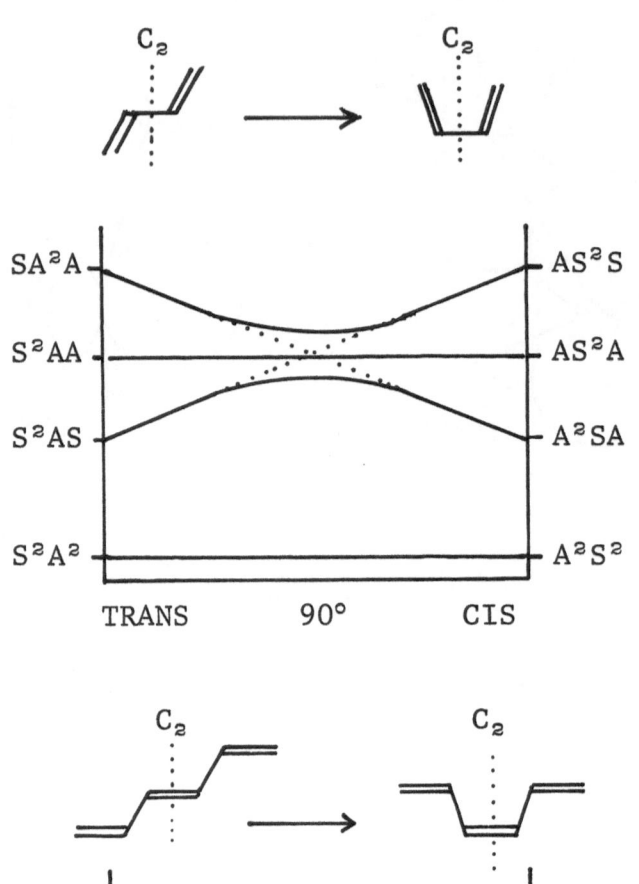

Figure 2

about the magnitude of the barrier or if the excited singlet or triplet states have barriers. Actual measurements (82) and calculations (58,59) show that both the excited singlet and triplets are nonplanar. The rules can be extended to the higher polyenes, butadiene, hexatriene (Figure 2) in which the symmetry behavior of the molecular orbitals of the trans, cis, cisoid or transoid configurations (with regard to the central bond of the system) is such that the final state diagrams for butadiene and hexatriene are vastly different. For twisting about the central single bond in butadiene the rules indicate a low barrier in the ground state but a high barrier in the excited state. Conversely the twisting about the central essential double bond in hexatriene yields a behavior similar to ethylene. Strictly the orbital symmetry rules can say nothing about twisting about the terminal double bond in butadiene. As previously mentioned in Rule E, actual calculations on butadiene (31) indicate that the triplet and ground state energies are nearly the same as the geometrical configuration of the molecule approaches the point where an "essential" double bond is twisted 90°. The only exception is when a zwitterionic intermediate is energetically favored (31). In the polyenes and other simply substituted ethylenes twisting about the double bond produces a potential energy surface reminiscent of ethylene. Thus the results of the orbital symmetry rules go beyond rigid symmetry considerations.

Classically a large amount of chemical information has been interpreted using resonance arguments. The computational support from valence bond calculations for classical resonance theory is weak. Attempts to rationalize the spectral features of organic molecules using the usual charge transfer type resonance structures are misleading (84-86). Thus, two important considerations should be considered when such attempts are made: i) there is no noncomputational method of assigning the relative importance of various resonance structures in the excited states of molecules; and ii) when attempting to rationalize the mechanisms of organic photochemical reactions using resonance theory one must include only biradical or biradical-ionic (31,86) structures if the reaction occurs through a triplet mechanism (unless intersystem crossing is invoked). To date the attempt to rationalize photochemical mechanisms using resonance structures have stemmed from the desire to correlate the experimental findings within the context of qualitative theoretical organic chemistry. The Woodward-Hoffmann Rules present an improvement over resonance theory in this respect. However, the Woodward-Hoffmann Rules also suffer from a lack of computational support from the actual calculation of the potential energy surfaces. A number of people are vigorously at work on these calculations and the prediction is that the Woodward-Hoffmann Rules will eventually have computational support. Perhaps the time will return when valence bond calculations or projections techniques applied to the molecular orbital wavefunctions will produce some support for resonance arguments.

Finally, it should be reemphasized that whatever the prediction of the Woodward-Hoffman rules or actual calculations on the potential energy surfaces, excited singlet and triplet molecules eventually return to the ground state. Thus, nonradiative processes certainly play an important role in photochemical processes. The potential energy surfaces are not the whole story.

Acknowledgement: This work was supported by the National Space Agency, NGR 05-061-004.

REFERENCES

1. G. Herzberg, "Spectra of Diatomic Molecules," Chapt. 3, D. Van Nostrand (1950).

2. _____, "Electronic Spectra of Polyatomic Molecules," Chapt. 3, D. Van Nostrand (1966).

3. Ref. 2, pp. 429-437.

4. J. R. McNesby and H. Okabe, "Advances in Photochemistry," Vol. 3, p. 185, Interscience (1964).

5. For a discussion see C. A. Coulson, "Reactivity of the Photo-excited Organic Molecule," pp. 1-49, Proc. of the 13th Conference of the Solvay Institute (1965), Interscience (1967).

6. D. Phillips, J. Lemaire, C. S. Burton, W. A. Noyes, Jr., "Advances in Photochemistry," Vol. 5, pp. 329-364, Wiley and Sons (1968).

7. R. A. Young, G. Black, and T. G. Slanger, J. Chem. Phys., $\underline{48}$, 2067 (1968).

8. For a discussion see J. G. Calver and J. N. Pitts, Jr., "Photo-chemistry," pp. 622-624, Wiley and Sons (1966).

9. For a recent review of ketone photochemistry see J. S. Swenton, J. Chem. Ed., $\underline{46}$, 217 (1969).

10. C. H. Bibart, M. C. Rockley and F. S. Wettack, J. Am. Chem. Soc., $\underline{91}$, 2802 (1969).

11. E. J. Baum, L. D. Hess, J. R. Wyatt and J. N. Pitts, Jr., ibid, $\underline{91}$, 2461 (1969).

12. H. L. Hyndman, B. M. Monroe, and G. S. Hammond, *ibid*, <u>91</u>, 2852 (1969)

13. J. Saltiel, K. R. Neuberger, and M. Wrighton, *ibid*, <u>91</u>, 3658 (1969).

14. N. E. Lee and E. K. C. Lee, J. Chem. Phys., <u>50</u>, 2094 (1969).

15. H. E. Zimmerman and R. W. Elser, J. Am. Chem. Soc., <u>91</u>, 887 (1969).

16. _____, and P. S. Mariano, *ibid*, <u>91</u>, 1718 (1969) and papers cited therein.

17. A. Padwa, W. Eisenhardt, R. Gruber and D. Pashayan, *ibid*, <u>91</u>, 1857 (1969).

18. J. N. Pitts, Jr., D. R. Burley, J. C. Mani and A. P. Broadbent, *ibid*, <u>90</u>, 5902 (1968).

19. H. E. Zimmerman and K. G. Hancok, *ibid*, <u>90</u>, 3749 (1968).

20. For a recent review see T. Förster, Angew, Chemie (Eng. Ed.), <u>8</u>, 333 (1969).

21. For a discussion of charge transfer triplets see S. P. McGlynn, T. Azumi, and M. Kinoshita, "Molecular Spectroscopy of the Triplet State," pp. 88-90, Prentice Hall (1969).

22. H. Morrison and R. Kleopfer, J. Am. Chem. Soc., 90, 5037 (1968).

23. P. J. Wagner and D. J. Buckeck, *ibid*, <u>90</u>, 6531 (1968).

24. H. Ymazaki and R. J. Cventanovic, *ibid*, <u>91</u>, 521 (1969).

25. N. J. Turro and P. A. Wriede, *ibid*, <u>91</u>, 6863 (1969).

26. For a review see E. L. Wehry and L. B. Rogers in "Fluorescence and Phosphorescence Analysis," ed. D. M. Hercules, pp. 125-135.

27. H. E. Zimmerman and R. R. Sandel, J. Am. Chem. Soc., 85, 915 (1963).

28. A Zweig, "Advances in Photochemistry, Vol. 6, pp. 425-449, Wiley and Sons (1969).

29. J. P. Paris, ref. 26, pp. 190-193.

30. H. S. Pilloff and A. C. Albrecht, J. Chem. Phys., <u>49</u>, 4891 (1968).

31. E. M. Evleth, Chem. Phys. Letters, 3, 122 (1969).

32. P. Borrell and H. H. Greenwood, Proc. Roy. Soc., (London) A298, 453 (1967).

33. K. Inuzuka and R. S. Becker, Nature, 219, 383 (1968).

34. B. S. Rabinovitch, J. E. Douglas and F. S. Loonly, J. Chem. Phys. 20, 1807 (1952).

35. S. W. Benson, "The Foundations of Chemical Kinetics," Table XI.3, p. 254, McGraw-Hill (1960).

36. The issue of the possible singlet route for photo-cis-trans-isomerization is not completely resolved for the butadienes and stilbenes. Experiments to date that it does occur, see Refs. 37 and 38.

37. R. Srinivasan, J. Am. Chem. Soc., 90, 4498 (1968).

38. J. Saltiel and E. D. Megarity, ibid, 91, 1265 (1969) and papers cited therein.

39. For a discussion see N. J. Turro, "Molecular Photochemistry," pp. 212-216, W. A. Benjamin (1967).

40. W. G. Dauben, Ref. 5, p. 174.

41. G. S. Hammond, Ref. 5, pp. 313-315.

42. W. Th. A. M. Van der Lugt and L. J. Oosterhoff, Chem. Comm., 1235 (1968).

43. G. M. J. Schmidt, Ref. 5, pp. 268-277.

44. W. F. Richey and R. S. Becker, J. Chem. Phys., 49, 2092 (1968).

45. A. G. Wahl, P. J. Bertoncini, G. Das, and T. L. Gilbert, Int. J. Quant. Chem. 1, 123 (1967).

46. B. Levy and G. Berthier, ibid, 2, 307 (1967).

47. A. Veillard and E. Clementi, Theoret. Chim. Acta (Berl.), 13, 7, 133 (1967).

48. R. J. Buenker and S. D. Peyerimoff, ibid, 12, 183 (1968).

49. R. C. Morrison and G. A. Gallup, J. Chem. Phys., 50, 1214 (1969).

50. W. A. Goddard, Phys. Rev., 157, 73, 81 (1967).

51. J. P. Malrieu, Photochem. and Photobiol., 7, 531 (1968).

52. N. L. Allinger and T. S. Stuart, J. Chem. Phys., 47, 4611 (1967) and papers cited therein.

53. E. M. Evleth, ibid., 46, 4151 (1967).

54. S. D. Peyerimoff and R. J. Buenker, ibid, 49, 2261 (1968).

55. R. J. Buenker and L. L. Whitten, ibid, 49, 5381 (1968).

56. S. D. Peyerimoff and R. J. Buenker, ibid, 49, 2473 (1968).

57. R. Hoffmann, ibid, 39, 1397 (1964) and subsequent papers.

58. R. J. Buenker, ibid, 48, 1368 (1968).

59. U. Kaldor and I. Shavitt, 48, 191 (1968).

60. D. T. Clark and D. R. Armstrong, Theoret. Chim. Acta, 13, 365 (1969).

61. J. J. McCullough, H. Ohorodnyk and D. P. Santry, Chem. Comm., 570 (1969).

62. G. Feler, Theoret. Chim. Acta, 12, 412 (1968).

63. A. C. Hopkinson, R. A. McClelland, K. Yates and I. G. Csizmadia, ibid, 13, 81 (1969).

64. A. S. N. Murthy, R. E. Davis, and C. N. R. Rao, ibid, 13, 81 (1969).

65. D. B. Chesnut and R. W. Moseley, ibid, 13, 230 (1969).

66. J. M. Lehn, B. Munsch, P. Millie, and A. Veillard, ibid, 13, 313 (1969).

67. K. Morokuma and L. Pedersen, J. Chem. Phys., 48, 3275 (1968).

68. E. Clementi and A. Clementi, ibid, 47, 3837 (1967).

69. J. A. Pople, D. P. Santry and G. A. Segal, ibid, 43, S129 (1965).

70. For a recent review see H. H. Jaffe, Acc. of Chem. Res., 2, 136 (1969).

71. For a general discussion of all semiempirical methods see
    M. J. S. Dewar in "Molecular Orbital Theory of Organic Chemis-
    try".

72. C. Giessner-Prettre and A. Pullman, Theoret. Chim. Acta, 13,
    265 (1969).

73. J. Del Bene and H. H. Jaffee, J. Chem. Phys., 48, 1807, 4050
    (1968).

74. More recent calculations including doubly excited configura-
    tions by these authors are in press.

75. M. A. Robb and I. G. Csizmadia, J. Chem. Phys., 50, 1819,
    (1969).

76. For a discussion of Walsh's rules see Chapter 3, Ref. 2.

77. H. C. Longuet-Higgins and E. . Abrahamson, J. Am. Chem. Soc.,
    2045 (1965).

78. R. B. Woodward and R. Hoffmann, ibid, 87, 395 (1965).

79. R. Hoffmann and R. B. Woodward, Acc. of Chem. Res., 1, 17
    (1968).

80. J. J. Volmer and K. L. Servis, J. Chem. Ed., 45, 214 (1968).

81. G. B. Gill, Quart. Rev., 22, 338 (1968).

82. Ref. 2, pp. 533-535.

83. E. M. Evleth, J. Am. Chem. Soc., 89, 6445 (1967).

84. H. E. Zimmerman, "Advances in Photochemistry," Vol. 1, pp.
    183-207 Interscience (1963).

85. Ref. 39, A discussion of the limitation of the use of such
    structures, pp. 162-170 and pp.237-241.

86. H. E. Zimmerman, R. W. Binkely, J. J. McCullough, and
    G. Zimmerman, J. Am. Chem. Soc., 89, 6589 (1967) and footnote
    42 of this same paper.

PHOTOELIMINATION REACTIONS OF MACROMOLECULES

R. F. Reinisch, H. R. Gloria, and G. M. Androes*

Ames Research Center, NASA

Moffett Field, California, 94035

INTRODUCTION

At the present time, our understanding of macromolecular photo-chemistry is guided by the recognition that the overall photolysis reaction is attributed to a combination of mechanisms each of which would have to be studied independently for a proper understanding of the photoelimination process. Isolation of each mechanism is often experimentally difficult to achieve, and the need often arises for an analytical method which can predict and explain the behavior of complex photochemical systems. Our initial[1] analytical description of the course of photoelimination reactions was concerned primarily with the rates of production and annihilation of electronically ex-cited intermediates, rather than with the energy content and elec-tronic configuration of the separate states. The basis for the earlier kinetic description was that the quantum yield of product formation, ϕ_{PC},[2] is given by the summation equation,

$$\phi_{PC} + \phi_T + \phi_I + \phi_L + \phi_{REV} = 1$$

where ϕ_T is the triplet yield, ϕ_I is the singlet and triplet radia-tionless relaxation yield, ϕ_L is the total luminescence yield, and ϕ_{REV} represents relaxation via reversible processes yielding no net photochemical reaction. The magnitude of each of these competing processes and the yield for product formation could be readily esti-mated by the kinetic analysis.[1]

*National Research Council Associate.

The purpose of this discussion is to present a simplified and concise, heuristic description, by means of energy level diagrams, of the possible pathways by which the primary excited intermediate must pass on its way to product(s). The detailed construction of these energy level diagrams will be described for photoelimination reactions in the condensed phase where vibrationally excited ground state intermediates are unimportant.[3] This important class of photolysis reaction is a special case of unimolecular bond scission that leads to the formation of two species: a large polyatomic radical; and a small radical fragment, e.g., a chlorine atom, an acetyl radical, or the simplest of free radicals, the electron. Depending upon the total excitation energy available, the polyatomic fragment can be born in its electronic ground state or can be electronically excited, while the small radical fragment will generally be in its electronic ground state. Once a primary radical is formed, its subsequent fate is independent of its mode of formation and is a function only of its intrinsic chemical reactivity and its environment. Therefore, this treatment of macromolecular photoelimination will focus attention on the energetics of primary intermediate formation with the goal of explaining the causes of the variation of photochemical behavior from one macromolecular system to another. Such a descriptive method permits the design and interpretation of photochemical experiments, and assures us that our observations are consistent with the predicted behavior of this general class of reactions.

The underlying purpose in presenting various examples of photoelimination reactions will be to demonstrate the generality and applicability of the energy level analysis to the explanation and prediction of the photochemical behavior of synthetic and natural macromolecules. One of the distinctive features of macromolecular photochemistry occurs in the sequential bonding of structural units containing the primary chromophores. This permanent spatial relationship between neighboring and distant chromophores is to be contrasted with the dynamic relationship between neighbors of a small monomolecular solute in a solvent. The pattern of sequential bonding is often a simple repeating one in the case of a typical synthetic polymer, while in natural polypeptides the amino acids are arrayed in a highly complex pattern along the peptide chain. This fixed geometry between residues may, for example, per-

mit energy transfer by dipole-dipole resonance resulting in a shift
of electronic energy from the primary chromophore to a distant acceptor. The examples to be presented in the biopolymer field are
monomeric aromatic amino acids. An understanding of the photochemical behavior of these isolated primary chromophores is required before we can extend the energy level method for predicting the behavior of polypeptides composed of a variety of aromatic amino acids
arrayed in a known spatial pattern.

ANALYSIS OF REACTION PATHWAYS

Absorption of a photon by organic macromolecules in the condensed state initiates a series of intramolecular processes that
partition the absorbed energy during the decay of the excited molecule to the ground state. Consideration of the energy levels of
reactive intermediates involved in the various relaxation processes
can provide information on the efficiency of each of the competing
processes, as well as give insight to the most probable pathway
that the photoelimination reaction should follow.[2,4] Quantitative
values for the energy levels of these reactive intermediate states
can be obtained from spectroscopic data. At the present time there
exists a large body of useful information about many macromolecules
of interest to polymer chemists and biochemists.

Energy Level Diagrams

For the present analysis, it has proved useful to classify
spectroscopic data in terms of the electronic configuration of the
excited states to be studied. Electron configuration determines
the singlet or triplet character of the state as well as its photochemical behavior. These data are used in conjunction with thermodynamic values of bond dissociation energies[2] and Hess Law summa-
tions[5] of reaction energetics to construct energy level diagrams of
the photochemical system. Figure 1 illustrates the possible energy
levels through which a hypothetical photochemical system passes
during the course of the reaction. The ordinate is in arbitrary
energy units and the abscissa represents progress towards products.
The shaded areas indicate unspecified vibrational energy levels associated with the various states.

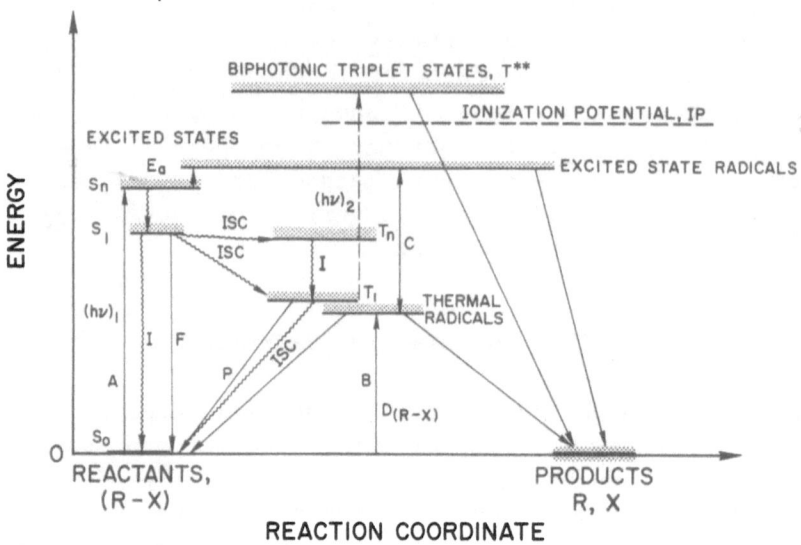

Fig. 1. - Generalized energy level diagram for photoelimination
         reactions of macromolecules.

    Direct excitation (Process A) of the reactant occurs at an
energy $(h\nu)_1$ characteristic of the reactant's chromophoric system
forming an excited singlet state S_n with energy E above the ground
state. In the condensed phase, the excited singlet can undergo internal conversion (Process I) or fluorescence emission (Process F)
from the lowest singlet state, S_1, and return to the ground state
S_0. In this analysis, the intersection of the absorption and emission spectra was used to locate the zero-zero band of S_1.[6a] If the
energy gap between singlet and triplet states is small, intersystem
crossing becomes more probable, and the excited singlet can undergo spin inversion (Process ISC) and enter the triplet manifold.
The lowest triplet state, T_1, can likewise undergo intersystem
crossing (Process ISC) or phosphorescence emission (Process P) and
return to S_0. These relaxation processes result in no net chemical
change, and for most macromolecules, internal conversion (I) is the
dominant mode for energy dissipation.[1] Formation of final products
(R, X) exemplifies the results of a typical unimolecular photoelimination reaction. Approach to products can be from three distinct
states: thermal doublet radicals; excited state doublet radicals,
and biphotonic triplet states, T**. Thermolysis (Process B) of the
reactant can occur if the thermal energy available is equal to or
greater than the bond dissociation energy $\left(D_{(R-X)}\right)$ required for
homolysis. In addition to these dissociation processes, photo-

ionization of T^{**} can occur <u>via</u> the triplet manifold by a biphotonic excitation process, if the total excitation energy, i.e.,
$\left(T_1 + (h\nu_2)\right)$ is greater than condensed phase ionization potential
(IP). This ionization process leads to the formation of a radical
cation ($\cdot\overset{+}{R}$-X) and a stabilized electron. A more detailed discussion of the possible photo-dissociation processes will be presented
in the following sections.

Primary Reactions

The minimum energy level from which a radical reaction can
occur <u>via</u> the singlet route is determined by the sum of the dissociation energy (B) and the electronic excitation energy (C) of the
radical intermediate. From this highly reactive, electronically
excited doublet state, downward cascade to products or thermal
radicals is possible. If the approach to the excited state radical
is from a singlet level in excess of the sum of processes (B) and
(C), then the reaction will proceed at no expenditure of activation
energy E_a; if the approach to the excited radical state is from a
level possessing insufficient energy as shown in Figure 1, then the
energy deficit (E_a) must be supplied by the thermal environment.
Similar arguments apply if intersystem crossing to the triplet
manifold is allowed. Reaction through an excited triplet state can
occur from a level T_1 having greater energy than the thermal radical as shown in Figure 1, with nil activation energy required.
Conversely, a small activation energy may be required for triplet
levels containing energy just slightly below that of the thermal
radical. If the activation energy is too large, then a biphotonic
process in which two successive quanta are absorbed, can provide
another route to photoproducts. This alternate route occurs by
absorption of a second quantum by the lowest excited triplet.
This doubly excited triplet state, T^{**}, can photoionize to give a
thermal radical cation and a stabilized electron.

Secondary Reaction

Secondary reactions directly attributable to radical formation
can also occur. If the excited state radical doublet possesses a
chromophore which can be re-excited by incident light, then the rè-
action can continue to a "secondary" photolysis product as well as

the primary product. Different secondary reactions can occur if
the lowest triplet level can be re-excited by incident light. A
doubly excited triplet state T** can be formed by a two-photon
process, which in the case of aromatic compounds has been observed
to lead to photoionization as well as photolysis. An additional
reaction can occur after irradiation ceases. This "dark" reaction
will continue as long as the thermal radicals formed contain sufficient energy to cleave a chemical bond. Examples of these reactions
will be discussed in greater detail in subsequent sections of this
paper.

Radical Energy Levels

Thermal radical energies (Process B) can be estimated using
the bond dissociation values tabulated by Calvert and Pitts,[2]
while energy levels of excited state radicals (Process C) are usually
derived from the electronic absorption spectra[6b] as obtained by
flash photolysis or pulse radiolysis of the reactants. A Hess Law[5]
summation using thermodynamic values[2] is required to establish the
thermodynamic exo- or endothermicity of the overall reaction and
locates the energy position of the products relative to the reactants. Ionization potentials can be estimated from gas phase data
corrected by subtracting the solvation energy[7] of the electron
(39 kcal/mole). The ionization potentials used in this analysis are
probably higher than actual because the contribution from the stabilization energy of the radical cation is not known.

Activation Energy

The activation energy, E_a, is defined in the sense of the
Arrhenius theory, and for condensed systems must be acquired from
the bulk thermal energy of the reaction system. Since spectroscopy
has shown that in condensed media the electronically excited molecule is in thermal equilibrium vibrationally with its environment
within a time which is short compared to the time required for
photochemical reactions,[4] the value of E_a indicates the minimum
energy which the reacting system must possess in order that the
electronically excited intermediate be transformed into product(s).
Although activation energies in the excited states are not generally
the complete cause of rate differences in photochemical reactions,
it is apparent that if a reaction is to occur in an excited state in
less than 10^{-9} second at room temperature, the activation energy

must be low (not more than several kcal/mole). The prediction of low values of E_a for photochemical reactions of macromolecules is qualitatively in agreement with many experimental observations.[4,8,9] Since the singlet photochemical reaction will <u>not</u> be observed if E_a exceeds several kcal, the estimates of E_a from energy level diagrams such as Figure 1 are useful for predicting the most probable route for a photochemical reaction to follow. However, caution must be exercised in employing values of E_a estimated from energy level diagrams for the calculation of reaction rate constants. The observed efficiency of the reaction may reflect differences in rates of radiationless processes, as well as rates of photochemical reactions. The magnitude of all competing processes as well as the quantum yield for product formation can best be estimated by a previously developed analysis.[1] These results combined with reaction pathways delineated by use of the energy level diagrams can provide a more complete description of photoelimination reactions for macromolecules.

APPLICATION OF THE METHOD

Photoelimination reactions of macromolecules can follow diverse pathways leading to product evolution. This discussion will center on examples that undergo homolytic cleavage leading to reactions <u>via</u> two-centered intermediates. Two of the synthetic polymers chosen have been studied in great detail experimentally and will be used to illustrate the utility of the present analysis in elucidating the reaction pathways even for highly complex systems. The photochemical reactions of three aromatic amino acids, tyrosine, tryptophan, and phenylalanine will also be discussed. These amino acid residues play a significant role in the photochemical behavior of biopolymers, acting not only as principal chromophores, but also appearing as direct participants in reactions involving enzyme inactivation and biological damage. (See discussion of A. D. McLaren in "A Brief History of the Photochemistry of Macromolecules" in this work.)

Photolysis of Poly(vinyl chloride)

It is generally agreed that the photolysis of PVC leads to the formation of conjugated unsaturation in the polymer backbone accompanied by the production of hydrogen chloride.[10-13] Since alkyl chlorides exhibit no absorption maxima at wavelengths above 200 nm,[14] the observed reaction threshold of 330 nm for PVC has been baffling.[1]

Although Boyer[11] and others [12] have suggested that polyenic unsaturation is the principal chromophore, spectroscopic evidence consistent with this suggestion has been obtained only recently.[13] This foreign unsaturation probably originates from thermal dehydrochlorination or disproportionation reactions during processing. The generalized reaction

$$\left(\text{CH}=\text{CH}\right)_n\text{CH} \sim \xrightarrow{h\nu} \sim \left(\text{CH}=\text{CH}\right)_{n+1} + \text{HCl}$$

has a quantum yield of 0.01 for HCl production at 254 nm[1] and no other gaseous product has been detected by mass spectrometry.[12] A Hess Law summation has shown the photoelimination reaction to be endothermic by 7-10 kcal/mole. The photoelimination of HCl can be derived from either an intramolecular or a radical process; experimental observations and an energy level diagram for PVC will be utilized to determine the most probable pathway and to propose reasonable mechanisms which best describe the photoelimination reaction.

Figure 2 shows an energy level diagram for PVC photolysis based on polyenic chromophores with the number of double bonds, (n), ranging from 1 to 5 and whose absorption maxima[10,12] lie within the observed photolysis reaction threshold wavelength (PT).[12] The lack of fluorescence for polyenes smaller than a pentaene necessitates the use of absorption maximum rather than the zero-zero band for assignment of S_1 levels. Excitation of PVC containing these chromophores produces singlet states possessing at least 88 kcal/mole above the ground state S_0. The direct formation of polyenyl triplets is highly forbidden and the only known triplets have energies of 106 kcal/mole, 82.2 kcal/mole, and 69 kcal/mole for the case of ethylene, butadiene and hexatriene, respectively. These levels have been assigned by oxygen perturbation experiments.[15] Intersystem crossing from singlet to triplet has never been observed for polyenes, as evidenced by their lack of phosphorescence, and can be explained on the basis of the large singlet-triplet energy gaps shown in Figure 2.

It is apparent that direct excitation of the polyenes in PVC must lead to reactions occurring predominantly via electronically excited singlet states. This conclusion is experimentally con-

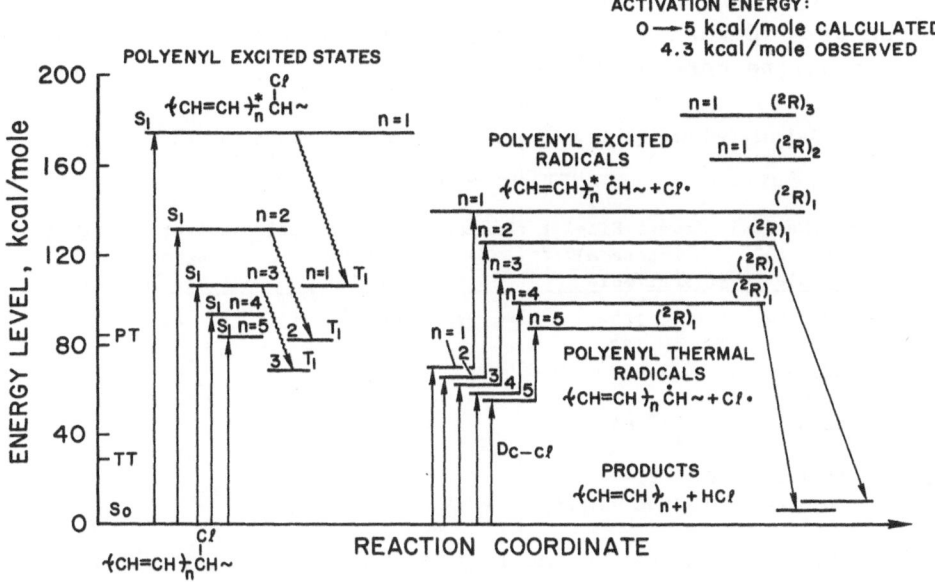

Fig. 2. - Energy level diagram for photolysis of poly(vinyl chloride).
Thresholds for photochemical (PT) and thermal (TT) reactions are marked on ordinate.

firmed by the inability of triplet quenchers to alter the quantum yield of hydrogen chloride production from PVC.[10]

Thermal dissociation of the C-Cl bond of a secondary alkyl chloride requires 82 kcal/mole[2] but the cleavage of a C-Cl bond conjugated to an ethylenic linkage is assisted by the exothermicity contributed by the formation of the resonance stabilized allyl radical. The resonance energy of the allyl radical is 12 kcal/mole[16] and locates the thermal radical from PVC at 70 kcal/mole. Benson[17] has shown that the resonance energy of the pentadienyl radical is 15.4 ± 1 kcal/mole. This value is only 25 percent larger than the allyl stabilization energy and a monotonic decrease of 3 kcal/mole was used as the basis for calculating the remaining trienyl, tetraenyl and pentaenyl radicals. Electronic excitation of the allyl radical produces an excited doublet state whose zero-zero band is located at 408 nm,[18] with maxima reported at 310 nm[19] and 258 nm,[20] respectively. Table I is a list of the absorption maxima estimated

for the dienyl, trienyl, tetraenyl, and pentaenyl radicals by sub-
traction of $\sim$ 10 kcal for each additional unsaturated linkage, as
suggested by the data of Bodily and Dole.[20]

TABLE I

Calculated Energy Levels for Conjugated Polyenes in PVC

$$H-\left(CH=CH\right)_n-H$$

$\underline{n}$	Thermal Radical (Process B)[a] kcal/mole	Lowest Singlet (Process A)[a,b] kcal/mole	Excited State Radical (Process C)[a] kcal/mole	Activation Energy $E_a = B+C-A$ kcal/mole
1	70	163	70	-24
2	66	132	60	-6
3	62.5	107	48.5	4
4	59.0	94	40	5
5	55.5	84	32	3.5

(a)   defined in Fig. 1.
(b)   absorption maximum values

Examination of Table I and Figure 2 shows that, while the
vinylene and diene require zero activation energy, the triene,
tetraene, and pentaene singlet excited states need an activation
energy of 4, 5, and 3.5 kcal/mole to reach the energy level of the
polyenyl excited radical and the ground state chlorine atom. The
reactive chlorine atom can rapidly abstract a nearby hydrogen atom
to give hydrogen chloride—a process requiring less than 1 kcal/mole
of activation energy.[21] Our studies have shown that the quantum
yield for hydrogen chloride formation from PVC is increased by temperature rise. Over the temperature range of -40 to +70°C, an
activation energy of 4.3 kcal/mole was measured.[22] This close
agreement between our experimentally observed value, and the value
predicted by the energy level diagram is consistent with the suggested singlet route.

The presence of photo-induced free radicals in PVC has been
detected by electron paramagnetic resonance spectroscopy. The results of our irradiation of oriented PVC films in vacuum and at low
temperatures are shown in Figure 3. At 82°K, a superposition of two
(or more) curves is observed: one component with a peak-to-peak
spread of $\sim$ 30 gauss shows no resolvable structure; the second component is broader and reveals some structure. Raising the temperature to 133°K produced a considerably altered spectrum whose integrated area and envelope shape was consistent with a transformation
of the low temperature species, as well as with a reduction in the
total number of unpaired electrons. On raising the temperature to
220°K, the spectrum continued to decrease in integrated area, and

displayed a narrower signal. Finally, at room temperature, the radical concentration is no longer detectable.

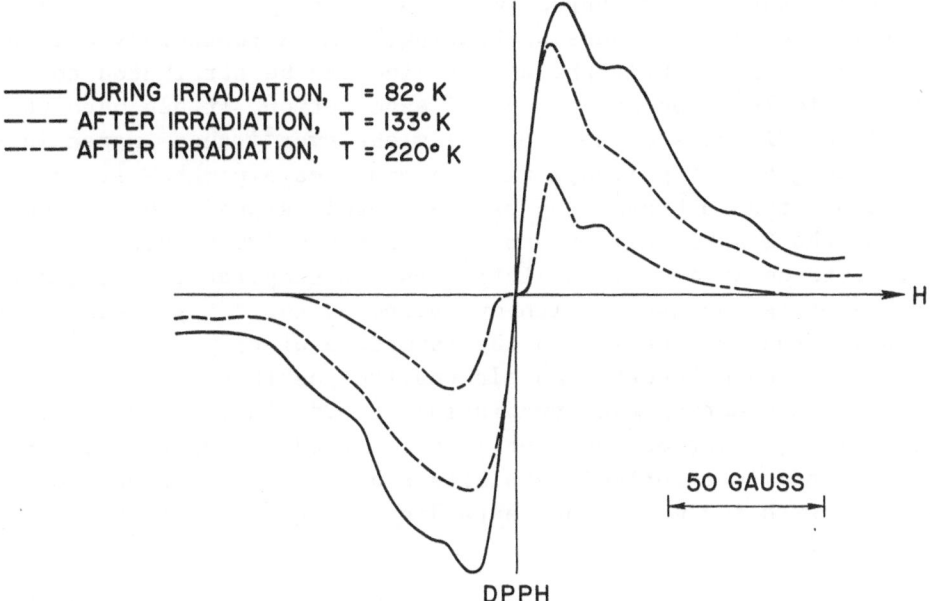

Fig. 3. - EPR spectra of free radicals in poly(vinyl chloride) during and after low temperature UV irradiation (g = 2 located with DPPH marker.)

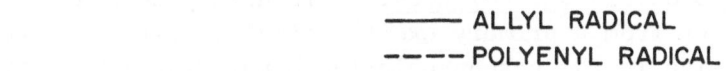

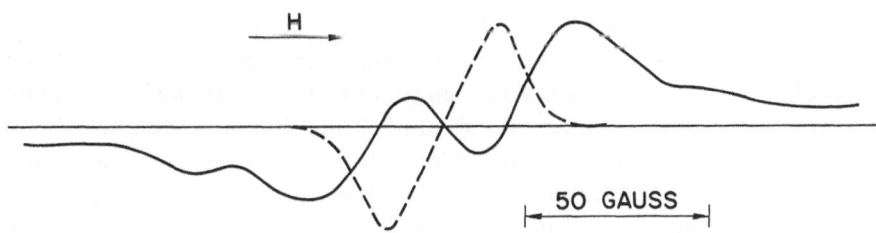

Fig. 4. - EPR spectra of radicals which disappear upon heating ir-radiated PVC. The sum of the two curves is equal to the difference between the $133^{\circ}K$ and $82^{\circ}K$ curves of Fig. 3.

Identification of the partially resolved spectrum seen at 82°K
was accomplished by a difference spectrum (Fig. 4) which indicated
a superposition of a single line and a quartet pattern with a splitting of $\sim$ 42 gauss and an approximate intensity ratio of $\sim$ 1:3:3:1.
The quartet, which disappears on heating, can be reasonably assigned
to the allyl radical while the single line can be attributed to
polyenyl radicals. Contrary to usual expectation, it has been theoretically predicted and experimentally observed[23] that increasing
the chain length of a polyenyl radical produces a progressive nar-
rowing rather than a broadening of the overall signal width. Comparison of the EPR signals obtained at 82°K and 220°K shows a
regular decrease in overall splitting as the temperature increases.
This behavior is consistent with our proposed secondary thermolysis
of the allyl radical leading to the formation of polyenyl radicals
of increasing chain length. An alternative possible origin of the
single line seen during PVC irradiation at 82°K is trapped chlorine
atoms. This explanation, however, can be ruled out because the EPR
spectrum of atomic chlorine[24] exhibits a broad structure asymmetric
about g = 2 with its main lines extending 100 gauss downfield and
50 gauss upfield.

The energy level diagram (Fig. 2) and the experimental findings
can now be used to formulate a scheme useful for describing the
path and mechanisms of the photolysis of PVC. The overall photolysis
reaction can be best understood as occurring through a sequence of
multi-step pathways involving short-lived intermediates.

Primary Photolysis via Polyenes. This homolytic bond fission
process can occur from a primary excited singlet state of a polyene
which is repulsive in the carbon-chlorine bond. The lone-pair
electrons on the chlorine atom can always interact effectively with
the π electrons of the polyene because the orbital containing the
lone pair is parallel to the π orbitals of the polyene. After dissociation of the C-Cl bond, the resulting unpaired electron then
participates in a three electron conjugation of the allyl radical.
The chromophore which is excited depends on the wavelength employed
in the experiment, but at 254 nm the principal chromophores would
consist of the trienic, dienic and vinylene groups[13,25] present in
the polymer backbone. The photoelimination process leading to the
formation of the excited allyl radical, longer polyenyl radicals,
as well as a ground state chlorine atom is:

I.

n = 1, 2, 3 . . .

This unimolecular mechanism involving the formation of a chlorine atom and a polyenyl radical is probably competing with a four-center photoelimination reaction yielding non-radical products, by reaction II.

II.

n = 1, 2, 3 . . .

Reaction II is of minor importance because the steric requirements of coplanarity for the concerted elimination mechanism might be expected to statistically preclude a large population of allylic chlorine atoms from participating in these photoelimination mechanisms. The proposed primary photolysis reaction (I) and the singlet energy levels of the polyenyl excited states shown in Figure 2 lead to an explanation for the threshold wavelength[12] of 330 nm. This wavelength corresponds to the energy level of the pentaenyl excited state. Any polyenyl excited state of conjugation longer than a pentaene does not possess sufficient energy to dissociate the C-Cl bond in the allylic position. However, direct excitation of any polyene located at energy levels above 88 kcal/mole leads to the production of the corresponding excited polyenyl radical and a chlorine atom as shown in reaction I.

It is apparent from the theoretically estimated[1] and experimentally observed activation energy of 4 kcal/mole that the electronically excited polyene is transformed into primary product(s) in $< 10^{-8}$ sec. The prompt formation of the allyl radical at 77°K as detected by EPR spectrometry is also consistent with the unimolecular radical mechanism.

Radical Initiation via Hydrogen Atom Abstraction. Production
of highly reactive chlorine atoms by process I leads to subsequent
reactions proceeding by hydrogen-atom abstraction from secondary,
tertiary, and/or allylic C-H bonds in the PVC chain, followed by
the subsequent formation of an allylic or an alkyl radical as well
as hydrogen chloride:

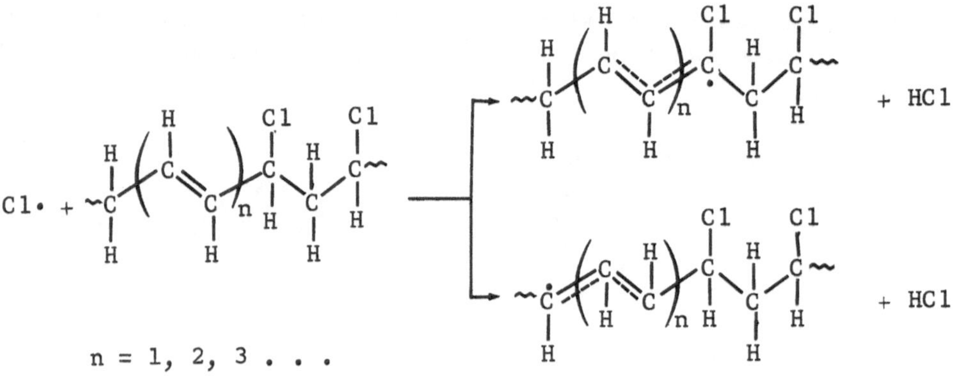

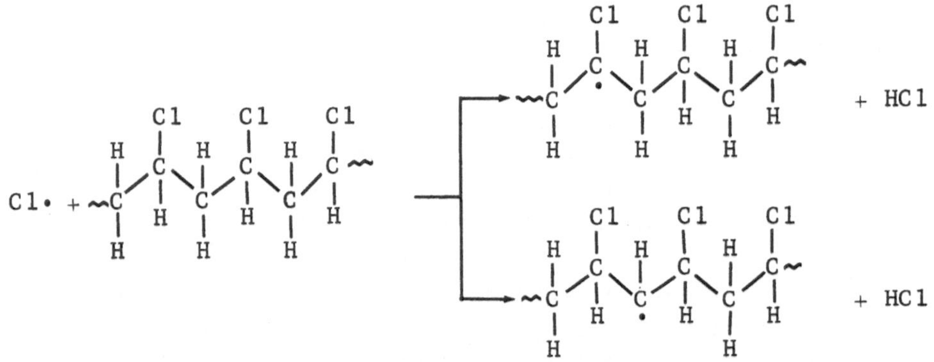

Although the reaction of chlorine atoms with hydrocarbons[21] is
known to occur at exceedingly rapid rates ($\sim 10^{11}$ sec^{-1}) with activation energies on the order of tenths of a kilocalorie, the reaction rate is non-selective with respect to the type of hydrogen
atom being attacked. In fact, the chlorine atom[26] abstracts tertiary hydrogens only 1.5 times faster than secondary hydrogens and
abstracts allylic hydrogens at about 0.7 times slower than secondary
hydrogens. Thus, the relative abundance of each radical will depend more on the initial population of each type of C-H bond rather
than on the kinetics of abstraction. Poutsma[26] has indicated that

the major factor controlling reactivity is the electron density of
the C-H bond being attacked by the highly electrophilic chlorine
atom. Finally, if the activation energy for hydrogen abstraction
is already as low as 1 kcal/mole,[21] it is apparent that molecular
or environmental modifications cannot produce any significant changes
in rates for a process that is already close to the diffusion controlled limit of $\sim 10^{11}$ M^{-1} sec^{-1}. This conclusion suggests that
the chlorine atom is too short-lived to be captured by any additive.

The formation of resonance stabilized allyl radicals via process III is consistent with the EPR evidence presented in Figures
3 and 4. The formation of highly reactive alkyl radicals via
process IV leads to thermal bimolecular reactions resulting in cross
linking or in secondary thermolysis reactions which have been observed as post-irradiation effects upon warming photolyzed PVC in
our EPR experiments.

Secondary Photolysis via Radicals. The polyenyl radicals produced by the initial reactions are new chromophores which can be
re-excited by the appropriate incident light and undergo secondary
photolysis reactions leading to the immediate production of a
conjugated biradical and a ground state chlorine atom:

V.

n = 1, 2, 3 . . .

The ground state chlorine atom generated by this secondary photolysis can participate in hydrogen abstraction reactions which cannot
be kinetically distinguished from those shown in processes (III) and
(IV). By contrast, the conjugated biradical formed in (V) will
promptly rearrange to form a polyene possessing one more unsaturated
linkage than its parent. Although the polyene formed in (V) cannot
undergo further photolysis because of the absence of intramolecular

allylic carbon-chlorine bonds, it can undergo further attack by
chlorine atoms. The net effect of the secondary photolysis reaction
(V) would be to reduce the steady state concentration of polyenyl
radicals generated by the primary photolysis. The observed satura-
tion of the EPR signal with increased irradiation time is consistent
with this proposed step.

Secondary Thermolysis via Radicals. The importance of secondary thermolysis reactions has been neglected in previous discussions
of the photolysis of PVC. On the other hand, in the radiolysis of
PVC, it has long been recognized that PVC, after irradiation by high
energy electron beams or X-rays in vacuum at -196°C[10] will spontaneously evolve hydrogen chloride at temperatures above -50°C.

Let us now consider what chemical reactions are energetically
possible after ultraviolet irradiation ceases. When illumination
is stopped, additional ground state reactions can occur deriving
their activation energy from the bulk thermal energy. As discussed
earlier, the EPR data can best be explained by the formation of
polyenyl radicals of increasing chain length by the intramolecular
elimination of hydrogen chloride as shown:

VI.

VII.

Bengough[27] and Wilson[28] have measured the activation energy for
thermodehydrochlorination of PVC and found a value of ~ 26 kcal/mole.
Using this value, it becomes apparent from Figure 2 that any polyenyl
thermal radical possessing energy in excess of 26 kcal/mole can undergo successive thermodehydrochlorination reactions. This stepwise
production of HCl will continue until each radical reaches an energy

less than 26 kcal/mole (level TT). Each thermal radical shown in
Figure 2 (ranging from n = 1 to 5) has sufficient energy to undergo
at least one dehydrochlorination step, with the allyl, dienyl, trienyl, and tetraenyl radicals capable of two successive steps. Spectroscopic evidence of this predicted limitation on the chain length
of conjugated unsaturation produced photolytically in PVC has been
obtained in our studies, where we have never detected polyenes with
unsaturation lengths greater than 6 or 7.[22] It can now be readily
understood why polyenyl unsaturation greater than six is never observed[22] in the photolysis of PVC while much longer extents of conjugation are observed upon warming after radiolysis[29,30] of the polymer. Ionizing radiation never reaches the energy limiting case of
ultraviolet photolysis, because the secondary electrons contain
more than sufficient energy to break a C-Cl bond at any location.
Although the photolysis and radiolysis of PVC both produce HCl, the
difference in extents of conjugation of the polyene products constitutes strong evidence that the primary mechanisms cannot be the same.
A more detailed discussion of this point will be discussed in a
forthcoming paper.[22]

The energy level diagram for PVC photolysis (Fig. 2) has provided valuable guidelines for the formulation of reaction schemes
that are consistent with experimentally observed facts obtained by
our studies as well as those of others.[10-13] Additional corrobora-
tion for the soundness of the energy level analysis is provided by
the close agreement with experimental values of E_a. The delineation
of the unique singlet route for photolysis reactions of PVC is in
excellent agreement with results of quenching studies in our laboratories.[22]

Photolysis of Poly(styrene)

Photolysis of poly(styrene) has been observed at 254 nm[9] and
the generalized reaction leads to formation of hydrogen as well as
chain unsaturation:

$$\left(CH_2 - \underset{\underset{\phi}{|}}{\overset{\overset{H}{|}}{C}}\right)_n \xrightarrow{h\nu} \left(CH = \underset{\underset{\phi}{|}}{C}\right)_n + H_2$$

For poly(styrene), our previous kinetic treatment for calculation
of quantum yields[1] gives a value for hydrogen production of 0.043
and an activation energy of 4.8 kcal/mole. These values are in good
agreement with observed values of 0.034 for the quantum yield for
H_2 and 2.9 kcal/mole for E_a.[9]

The energy level diagram for poly(styrene) is shown in Figure
5 with an excitation occurring at wavelengths between 190 and 273 nm.
Direct excitation of the principal chromophore at 207 nm produces
singlet level S_2[31] corresponding to 138 kcal/mole while the S_1 level
is located at 105 kcal/mole by the crossing point of the absorption
and fluorescence spectra.[32] Following the absorption process, the
formation of excimers (acronym for an excited state dimer) in poly-
(styrene) is allowed by the geometry of the molecule,[33] and the
excited singlet levels can be deactivated in 10^{-11} sec to the lowest dimer level S_E. The energy level of S_E has been established by
fluorescence emission from the excimer at 333 nm.[32] The lowest
triplet level for poly(styrene) is at 85 kcal/mole (335 nm) as
located by magneto-optic experiments.[34] This value is in good
agreement with the reported[35] triplet absorption at 347 nm for an
O_2-perturbed crystalline sample of toluene. The small value of the
singlet-triplet ($S_E \rightarrow T_1$) splitting can lead to rapid population of
triplet levels for poly(styrene).

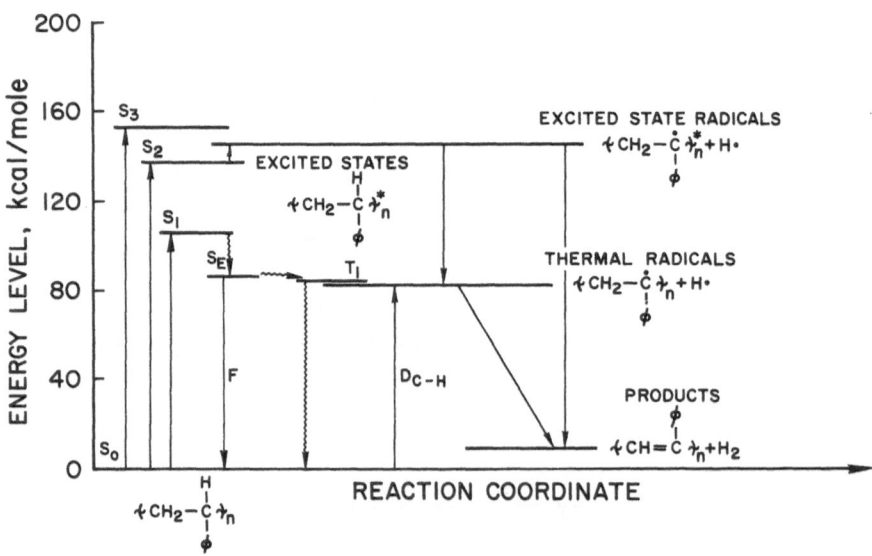

Fig. 5. - Energy level diagram for photolysis of poly(styrene).

As stated previously, photoelimination processes can proceed via two routes, either through excited state radicals or through triplets. For poly(styrene), the thermal dissociation of the weakest C-H bond produces a benzyl type radical with 83 kcal/mole.[2] Electronic excitation of this radical gives a zero-zero band at 62 kcal/mole (464 nm), based on the observed absorption[38] and emission spectra.[37] The lowest level for this electronically excited doublet state radical is then located by our summation method at 145 kcal/mole relative to ground state. The small energy gap observed for the S_2 level and the excess energy for the S_3 level suggest that the singlet photolysis route becomes significant only at wavelengths below 210 nm. Since reactions have been observed at 254 nm,[9] the photolysis of poly(styrene) must proceed via the triplet route. Intersystem crossing has a high probability as indicated from the energy level diagram, (Fig. 5) and the presumably repulsive triplet state decays with little or no activation energy to the thermal benzyl radical level. Subsequent formation of H_2 and unsaturation complete the reaction. Activation energies measured experimentally[9] (2.9 kcal/mole) are in accord with the triplet route prediction of the analysis, and the observed reaction threshold at 360 nm is in good agreement with the predicted threshold required by the benzyl thermal radical (345 nm). The estimated activation energy required for the longest wavelength reaction would be on the order of ~ 4 kcal/mole which is again in excellent agreement with observed values. The lack of large E_a requirement precludes the need for energetic biphotonic processes to explain poly(styrene) photolysis, which is in agreement with the measured unity intensity dependence of the reaction.[9]

Photo-Fries Rearrangement of p-Methoxyphenylacetate

The photo-Fries rearrangement is a condensed phase reaction exemplified by formation of substituted aromatic ketones from the photolysis of the esters derived from substituted phenols and aliphatic or aromatic carboxylic acids.[38] Upon absorption of light, cleavage of the ester C-O bond occurs with subsequent photoelimination of an acetyl radical and a phenoxyl radical. The final products are derived either by the reattachment of the acetyl radical at an ortho position, or by the loss of the acetyl group:

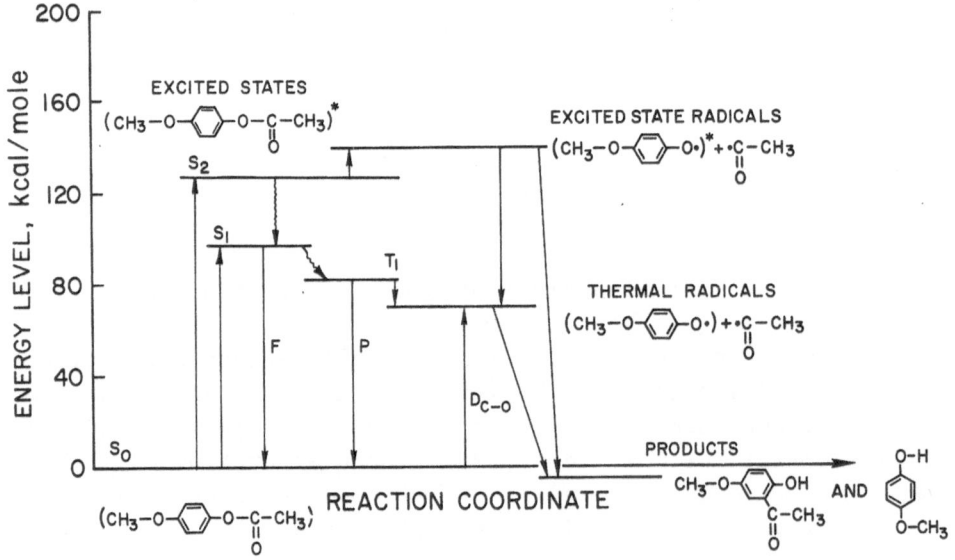

In the case of p-methoxyphenylacetate, the reaction takes place
by dissociation of the ester C-O bond, as established by the side
formation of p-methoxyphenol. Tracer experiments with ^{14}C carboxyl
labels have shown no isotope effect[39] which is consistent with little
or no activation energy required for the reaction. The intramolecular nature of this reaction has been postulated[40] to account for the
lack of interchange products in the analogous reaction of catechol
monoacetate and phenol irradiated in solution. Kobsa,[41] utilizing
the Hammet equation, has shown that a radical mechanism is most
probable for the rearrangement reaction. These observations are in
accord with the predictions that can be made by means of the energy
level diagram of Figure 6.

Fig. 6. - Energy level diagram for photo-Fries reaction of
        p-methoxyphenylacetate.

By analogy with hydroquinone, direct excitation of the aryloxy moiety of p-methoxyphenylacetate occurs at 293 nm and 225 nm[42] to produce singlets S_1 and S_2, possessing 97 and 127 kcal/mole, respectively. The S_1 level is in good agreement with the zero-zero band for hydroquinone reported by Stevenson.[43]

The thermal dissociation of an ether C-O bond requires about 79 kcal/mole.[2] This value is reduced by the 8 kcal/mole contributed by the resonance energy of the phenoxyl radical. By analogy with hydroquinone, electronic excitation of this para-substituted phenoxyl radical produces an excited state whose zero-zero band was located at 414 nm (69 kcal/mole) by flash photolysis.[44] The lowest level for this electronically excited phenoxyl radical is then located at 140 kcal/mole relative to ground state reactants.

Following the absorption process, the excited singlet states can undergo direct photochemistry or intersystem crossing to practically isoenergetic triplet levels. In p-methoxyphenylacetate, the lowest triplet level is at about 83 kcal/mole by analogy with the triplet of phenol.[45] This energy gap of 15 kcal/mole for singlet-triplet $(S_1 \rightarrow T_1)$ splitting makes less probable rapid rates of intersystem crossing. Alternatively, the singlet route even for the smallest energy interval from S_2 to the excited state radical would require an activation energy of 13 kcal/mole.

Since 300 nm irradiation of p-methoxyphenylacetate has been reported[38] to produce 2-hydroxy-5-methoxyacetophenone and p-methoxyphenol with nil activation energy, it seems reasonable that the photoproducts are formed with a repulsive triplet intermediate located at ~ 10 kcal/mole above the thermal radical. Population of this required intermediate would follow from the low but finite rates of intersystem crossing. Approach to the (doublet) thermal radicals from the lowest triplet state would be expected to occur with negligible activation energy. If the phenoxyl and acetoxyl radicals recombine before diffusion from the solvent cage, then either the rearrangement products or starting material will be formed. This primary recombination process occurs within 10^{-11} sec of the formation of the radicals,[3] before they are more than a molecular diameter apart. If the radicals have separated somewhat, and escape from the cage, the phenoxyl radical can abstract a hydrogen atom from a solvent molecule. Secondary reaction with solvent is consistent with the formation of p-methoxyphenol.

Photolysis of Aromatic Amino Acids

<u>Photolysis of Tyrosine</u>.   Observed photolysis reactions of
tyrosine at room temperature[4,6] proceed <u>via</u> energy levels shown in
Figure 7.   The excited singlet levels ranging from S_2 at 282 nm to
S_4 at 193 nm are assigned from absorption spectra.[47]   The location
of the zero-zero band at 99 kcal/mole (288 nm) was assigned by
using the crossing point of the absorption and fluorescence spectra
of <u>p</u>-cresol.[32]   The appearance of phosphorescence at 350 nm[45] arises
from the lowest excited triplet level, T_1, at 81.8 kcal/mole.
However, population of the triplet states from singlets formed by
direct excitation appears to be of low probability because the
singlet-triplet ($S_1 \rightarrow T_1$) splitting of 17 kcal/mole.   The low
probability of this intersystem crossing as well as rapid non-radiative transitions to S_0, accounts for the value[48] of $\sim 10^{-3}$ for
the quantum yield for phosphorescence.   Nevertheless, it has been
pointed out[49] that intersystem crossing from excited singlet states
other than the lowest singlet into the triplet manifold must occur
in condensed media.   The requirement for this alternate isoenergetic
process is that the crossing time must be of the order of magnitude
of 10^{-11} sec.   This means that the rate of isoenergetic intersystem
crossing must be equal to or greater than the rate of internal
conversion through the singlet manifold.

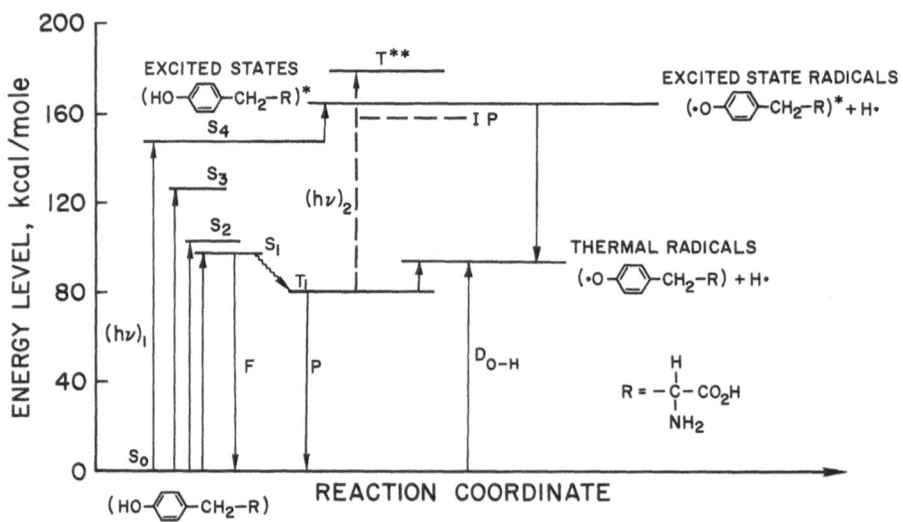

Fig. 7. - Energy level diagram for photolysis of tyrosine.

The absorption spectrum[50] of the excited radical doublet generated by the photolysis of p-cresol was used to calculate the energy levels of the excited radicals expected from the photolysis of tyrosine. Dobson and Grossweiner[49] observed maxima at 405, 388, and 358 nm corresponding to energies of 70.6, 73.7, and 79 kcal/mole. These values were added to the estimated O-H bond dissociation energy of 94 kcal/mole based on the difference between the 102 kcal/mole[2] required for alcohol O-H bond cleavage and the 8 kcal/mole gained in resonance energy by transforming phenol to phenoxide anion. We treat the phenoxyl radical as a phenoxide anion. To a first approximation the perturbation caused by the presence of a single unpaired electron rather than a pair of electrons in the nonbonding molecular orbital should have little effect on the change in resonance energy relative to phenol. Pauling[51] has indicated that the resonance energy of the phenoxide anion is about 8 kcal/mole greater than that for phenol. On this basis, the energy levels for the tyrosyl excited radical doublets were calculated to be 165, 168, and 173 kcal/mole.

It is apparent from inspection of Figure 7, that the large energy gap of 17 kcal/mole between the highest excited singlet state and even the lowest excited state radical doublet precludes photolysis reactions occurring through the singlet manifold. Despite the low probability of population of the triplet manifold, as discussed earlier, the observed photolysis of tyrosine at 254 nm must involve triplet intermediates. Since the lowest triplet contains insufficient energy to dissociate the tyrosyl O-H bond, the triplet mechanism must of necessity involve a third process for photolysis involving photoionization which leads to the formation of a radical cation and a solvated electron. Experimental evidence for the photoionization of tyrosine has been established by the following processes: (a) the analogous photoionization of phenol;[52] (b) irradiation, producing the absorption spectra of trapped electrons at 600 nm;[53] and (c) the irradiation of tyrosyltryosine, causing ejection of a photoelectron which was scavenged by monochloroacetic acid.[54] A two-photon scheme is proposed for the photoionization process, and from kinetic reasoning, the lifetime of the triplet intermediate is sufficiently long for absorption of a second quanta. The wavelength of the second photon absorbed in the process, $(h\nu)_2$, can either be identical to the incident wavelengths $(h\nu)_1$ or substantially different. In either case, a highly excited biphotonic triplet T** can be formed with sufficient energy to achieve photoionization as shown in Figure 7. The ionization potential level, IP, for tyrosine can be estimated from gas phase data[55] for phenol

(196 kcal/mole) less the solvation energy of the electron (39 kcal/
mole) (a partial correction applied to compensate for a photoionization process occurring in polar condensed media). Thus, only
about 157 kcal/mole are estimated to be required for the ejection
of a photoelectron from tyrosine. While the biphotonic character
of the photoionization rests upon experimental evidence for the
quadratic intensity dependence of the yield of stabilized electrons
from tyrosine, the similarity in the decay times of the intermediate
photoproduct and those reported for the triplet state of tyrosine,[53]
i.e., 1.3 and 0.95 sec, constitutes convincing kinetic evidence for
the biphotonic process' occurring through the triplet manifold.

The interpretation of the photochemistry of tyrosine is now
apparent. At 288 nm, the excited singlet state produced in the
primary step possesses only $\sim$ 100 kcal/mole and cannot undergo
photolysis. Isoenergetic intersystem crossing to the triplet manifold, followed by absorption of a second quanta by the lowest
triplet level T_1 would produce a highly excited triplet state T**
whose energy level would be equal to or greater than IP. Photoionization via T** then occurs, forming a stabilized electron and a
radical cation derived from tyrosine. Depending on the environment,
dissociation of the phenolic hydroxyl group of the radical cation
could lead to the phenoxyl radical and a proton. The reaction
scheme is:

$$^1(TYR) \xrightarrow{(h\nu)_1} {}^1(TYR)*$$

$$^1(TYR)* \xrightarrow{ISC} {}^3(TYR)*$$

$$^3(TYR)* \xrightarrow{(h\nu)_2} {}^3(TYR)**$$

$$^3(TYR)** \longrightarrow {}^2(TYR)^+_\bullet + e^-$$

If the radical cation and the electron recombine, no net photochemistry occurs. However, in a dipeptide, the stabilized electron has
been observed to move along the amide linkage and ultimately to be
captured by a vicinal sulfur atom in a cystinyl residue[56] or by a
C-terminal amino acid residue.[54] After electron capture, dissociation can occur leading to formation of an anion and a free radical.
This secondary dissociation process is analogous to atom abstraction
by a free radical.

Photolysis of Tryptophan. The photolysis of tryptophan at
254 nm has a quantum yield[46] of 2 $\times$ 10^{-2} at room temperature. The

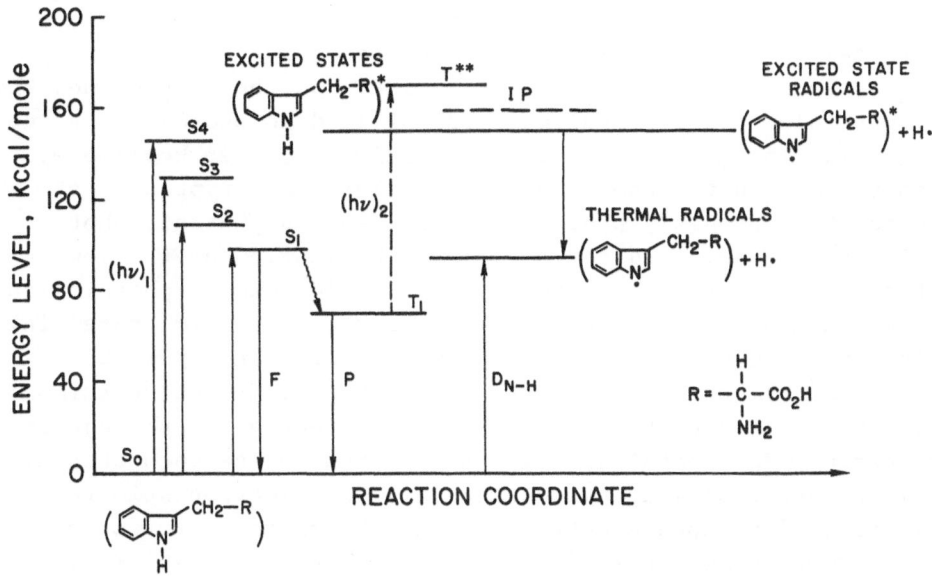

Fig. 8. - Energy level diagram for photolysis of tryptophan.

energy levels involved are displayed in Figure 8 which shows absorption occurring at singlet levels ranging from 269 nm, S_2, to 196 nm, S_4.[47] The lowest singlet level at 294 nm is assigned on the basis of the crossing point of the absorption and emission spectra of indole.[32] From phosphorescence measurements reported[47] and confirmed in this laboratory, the lowest triplet T_1 is located at 70.5 kcal/mole corresponding to 404 nm.[45] Since the probability of intersystem crossing varies as the square of the inverse of the singlet-triplet splitting,[57] the value of 17 kcal/mole makes this crossing less probable, and is a factor contributing to the estimated value of $\sim 10^{-3}$ for the quantum yield of phosphorescence.[48]

The absorption spectrum[58] of the excited radical produced by photolysis of tryptophan was required to locate the energy level of the tryptophyl radical. Grossweiner and Mulac[58] report the absorption maximum for the tryptophyl radical at 515 nm corresponding to 55.5 kcal/mole. This value, added to the estimated dissociation energy[2] for the N-H bond of the indole ring (95 kcal/mole), gives. a value of 150.5 kcal/mole for the electronically excited tryptophyl radical relative to the ground state reactants. The large energy

gap between all singlet levels below S_4 (196 nm) and the excited
state radical level precludes the singlet route for tryptophan
photolysis except in the vacuum ultraviolet. On the other hand, the
low lying triplet at ~ 70 kcal/mole cannot lead to photolysis since
the activation energy required is prohibitive (e.g. 25 kcal/mole).
Based on these results, the observed photolysis of tryptophan at
room temperature probably proceeds via the same biphotonic photoionization route described for tyrosine. Excitation of the lowest
triplet state T_1 by a second photon $(h\nu)_2$ would lead to photoionization if T** has sufficient energy to reach the ionization potential
level for tryptophan (I.P.). The level indicated in Figure 8
(~ 159 kcal/mole) has been estimated from theoretical calculations[59]
and corrected to compensate for the solvation energy of the ejected
photoelectron. The photoionization of tryptophan has been established
by observing the following processes: (a) the electron spin resonance spectra of trapped electrons;[60] (b) the photoionization of
analogous aromatic molecules like toluene, phenol, and aniline;[52]
(c) irradiation, producing the absorption spectra of trapped electrons at 600 nm.[53] The biphotonic character of the photoionization
rests upon experimental evidence for the quadratic intensity dependence of the yield of stabilized electrons. Lifetime measurements of the intermediate, T_1, are equal to those of the phosphorescence decay time of tryptophan,[53,60] e.g., ~ 6 sec.

   These experimental observations and the energy level diagram of
Figure 8 provide a basis for understanding some aspects of the
photolysis of tryptophan. Photoionization produces a stabilized
electron and the tryptophyl radical cation. Depending on the environment, the radical cation and the electron can recombine or the
electron might be captured by some reducible species such as a
neighboring cystinyl residue in a protein.

   Photolysis of Phenylalanine. The electronic aspects of the
previous analysis of poly(styrene) are directly applicable to the
photochemical behavior of phenylalanine. Excitation levels of
phenylalanine are shown in Figure 9, and correspond to observed absorption spectra maxima[47,61] ranging from 186 nm (S_3) to 207 nm (S_2).
The S_1 level is located at 273 nm by the crossing point of the absorption and emission spectrum and is based on data from toluene.[32]
In contrast to poly(styrene), poly-L-phenylalanine would not be expected to exhibit excimer emission because the large ring-ring
separation (~ 5Å) between alternating residues prohibits the formation of sandwich excimers. After absorption, intersystem crossing

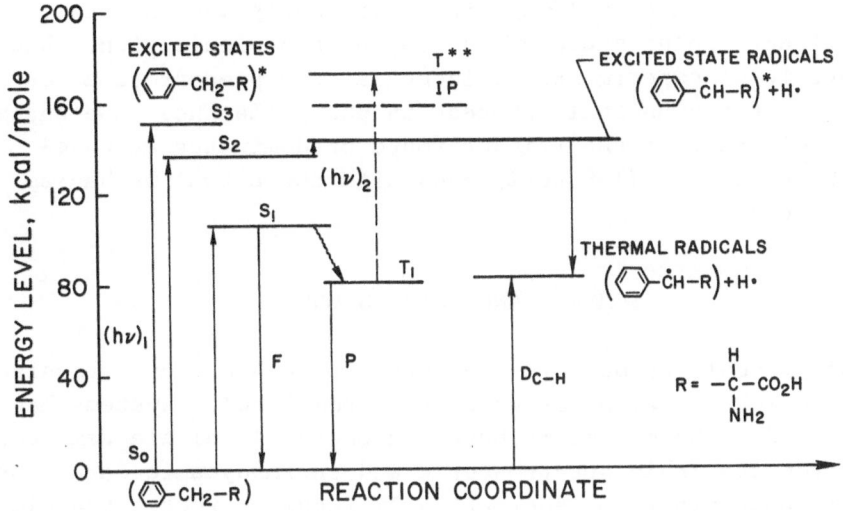

Fig. 9. - Energy level diagram for photolysis of phenylalanine.

can occur to populate the lowest triplet level T_1 at 82 kcal/mole located by observed phosphorescence at 347 nm.[45]

As in the case of poly(styrene), the excited benzyl radical (formed by thermal dissociation of the C-H bond[2] followed by electronic excitation)[36] is located at 145 kcal/mole relative to S_0. Singlet level S_2 requires 7 kcal/mole activation energy while singlet level S_3 requires no activation energy. The energy gap between the lowest singlet state and the excited radical doublet is 40 kcal/mole, precluding any significant contribution of the singlet route to photolysis reactions observed at 254 nm at room temperature.[46] The singlet-triplet energy gap of 22.5 kcal/mole precludes the population of the T_1 state by any route other than isoenergetic intersystem crossing as discussed earlier. Photolysis could proceed via the triplet route from T_1 with essentially zero activation energy.

The observed photolysis of phenylalanine at room temperature[46] could also proceed in part _via_ biphotonic photoionization routes described earlier. The mechanism is shown in Figure 9 with the biphotonic triplet state T** excited to an energy level sufficient to allow photoionization. The ionization potential shown in Figure 9 is based on gas phase data[55,63] for toluene (203 kcal/mole) reduced by the solvation energy of the photoelectron (39 kcal/mole). Despite the over-simplification of this calculation for phenylala-

nine, the level of 164 kcal/mole is sufficiently low that the
highly excited triplet state could lead to photoionization. However, there is no experimental evidence at the present time for the
occurrence of the biphotonic process in phenylalanine. The analysis
can only indicate that the triplet route predominates at wavelengths above 210 nm, (136 kcal/mole) with the actual mechanism
yet to be decided.

SUMMARY AND CONCLUSIONS

An analytical method which can predict, explain and organize
the photochemical behavior of complex macromolecular systems has
been presented. The method is based on energy level diagrams con-
structed from available spectroscopic and thermodynamic data. By
the use of these diagrams, energetically feasible routes traveled
by an electronically excited molecule to products can be predicted.
The descriptive method has provided an explanation for the causes
of the variation in photochemical behavior from one macromolecular
system to another, and has revealed some possibilities for the design of novel photostable structural materials. An immediate
practical result of the application of the method would be to reduce the amount of experimentation required for the solution of
photochemical problems and to predict behavior when experimental
data is lacking. In general, the refinement of thermodynamic and
spectroscopic data would not alter the overall picture presented by
the energy level diagrams.

Perhaps the most straightforward application of energy level
diagrams to macromolecular photochemistry is the analysis of photochemical systems that are complicated by secondary thermolysis and
photolysis reactions. In the case of the photolysis of poly(vinyl
chloride), the significant role of the subsequent thermolysis and
photolysis reactions, as well as the value of the activation energy,
the observed distribution of polyenes, and the singlet character of
the reaction were all explicitly predicted by the energy analysis.
In the photolysis of poly(styrene) excimer formation serves to reduce the singlet-triplet splitting, making the triplet route the
dominant one. By contrast with the photolysis of poly(styrene),
although the photo-Fries reaction is also triplet in character, the
energy level diagram indicates that the transitory intermediate is
a repulsive triplet decaying with no activation energy to the thermal radical. This variation in photochemical behavior from one system to another is simply explained by the energy level diagram.

Inspection of the ordering of states in the singlet manifolds for the examples presented shows that, for wavelengths below about 220 nm, the singlet route will dominate for photoelimination reactions because the energy gap between the spectroscopic singlet level and the electronically excited radical progressively decreases while singlet-triplet splitting ($S_n \rightarrow T_1$) is too large for rapid rates of intersystem crossing. In the case of poly(vinyl chloride), one concludes from the analytical method that the allyl radical will be the dominant intermediate in the vacuum ultraviolet as well as in the conventional ultraviolet region.

Several important conclusions have resulted from the application of the energy level analysis to photoionization processes in the aromatic amino acid residues of biopolymers. Although the principle of photoreversibility has been suggested as an energy dissipative process for organic molecules, little attention has been given to the concept of reversible photoionization as a mechanism for protecting polymers from ultraviolet degradation. If the efficiency of energy transfer between aromatic amino acids in a biopolymer permits the photoionization process, then the electron can be readily captured by the positive "hole" of the radical cation and no net photochemistry will occur. Adaptation of this concept of photoelectron recombination to synthetic polymers opens up the possibility of creating a new class of photostable polymers. Another conclusion is that a re-examination of the results of the ultraviolet irradiation of aromatic amino acids and biopolymers may be necessary because most previous studies did not consider the possibility of biphotonic photoionization process producing a radical cation and a stabilized electron. Successful elucidation of these photoionization processes may have a strong bearing on primary processes in radiation chemistry where ionization processes are abundant.

Although the emphasis throughout this article is on aspects from the authors' studies, the detailed construction of energy level diagrams can be applied to solving photochemical problems in regions of chemistry, biology, medicine, engineering and physics.

References

1.   R. F. Reinisch and H. R. Gloria, A.C.S. Polymer Preprints 9, 349 (1968).

2.   J. G. Calvert and J. N. Pitts, Jr., "Photochemistry," John
     Wiley & Sons, New York, N. Y., 1966, Chaps. 4, 6 and appendix.

3.   R. M. Noyes, in "Progress in Reaction Kinetics," G. Porter, ed.,
     Vol. 1, Pergamon Press, N. Y. 1961, p. 129.

4.   J. N. Murrell, "Theory of Electronic Spectra of Organic Mole-
     cules," Methuen, London, 1965.

5.   S. Glasstone, "Textbook of Physical Chemistry," D. Van Nostrand
     Co., Inc., New York, N. Y., 1946, p. 204.

6a. Private communication from E. M. Evleth.

6b. A. Habersbergerova, "Ultraviolet and Visible Spectra of Elec-
     trons, Unstable Radicals and Ions." English Translation in
     U.S.A.E.C. Facsimile Report No. UJV-1625.

7.   M. Anbar and E. J. Hart, J. Phys. Chem. 69, 1244 (1965).

8.   F. H. Winslow, W. Matreyek, A. M. Trozzolo and R. H. Hansen,
     ACS Polymer Preprints 9, 377 (1968).

9.   N. Grassie and N. A. Weir, J. Appl. Poly. Sci. 9, 975 (1965).

10. W. C. Geddes, Rubber Chem. Technol. 40, 178 (1967), see also
     ref. 22.

11. R. F. Boyer, J. Phys. Colloid Chem. 51, 80 (1947).

12. R. F. Reinisch, H. R. Gloria, and D. E. Wilson, ACS Polymer
     Preprints 7, 372, (1966).

13. R. F. Reinisch and H. R. Gloria, J. Solar Energy 12, 75 (1968).

14. G. Herzberg, "Electronic Spectra of Polyatomic Molecules,"
     D. Van Nostrand Co., Inc., Princeton, N. J., 1967, p. 541.

15. D. F. Evans, J. Chem. Soc. (1960), 1735.

16. D. M. Golden, N. A. Gac, and S. W. Benson, Abstracts, 157th
     National Meeting, Amer. Chem. Soc. April, 1969.

17. K. W. Egger and S. W. Benson, J. Am. Chem. Soc. 88, 241 (1965).

18. C. L. Currie and D. A. Ramsay, J. Chem. Phys. _45_, 488 (1966).

19. E. J. Burrel, Jr. and P. K. Battacharyya, J. Chem. Phys. _71_, 774 (1967).

20. D. M. Bodily and M. Dole, J. Chem. Phys. _45_, 1428 (1966).

21. A. F. Trottman-Dickenson in "Advances in Free Radical Chemistry, Vol. I, G. H. Williams, Ed., (Logos), London, 1965, p. 11.

22. R. F. Reinisch, H. R. Gloria, and G. M. Androes, to be published.

23. M. W. Hanna, et al. J. Chem. Phys. _37_, 361 (1962).

24. C. L. Gardner, J. Chem. Phys. _46_, 2991, (1967).

25. F. Sondheimer, et al. J. Am. Chem. Soc. _83_, 1675 (1961).

26. M. L. Poutsma, J. Org. Chem. _31_, 4167 (1966).

27. W. I. Bengough and I. K. Varma, Europ. Poly. J. _2_, 49 (1966).

28. D. E. Wilson and F. M. Hamaker, in "Thermal Analysis," Vol. I, R. F. Schwenker and P. Garn, Eds., Academic Press, Inc., New York, N. Y., 1969.

29. S. Ohnishi, Y. Nakajima, and I. Nitta, J. Appl. Polymer Sci. _6_, 629 (1962).

30. E. J. Lawton and J. S. Balwit, J. Phys. Chem. _65_, 815 (1961).

31. H. H. Jaffe and M. Orchin, "Theory and Applications of Ultraviolet Spectroscopy," John Wiley and Sons, Inc., New York, N. Y., 1962, p. 257.

32. I. B. Berlman, "Handbook of Fluorescence Spectra of Aromatic Molecules," Academic Press, New York, N. Y. 1965.

33. M. T. Vala, J. Haebig and S. A. Rice, J. Chem. Phys. _43_, 886 (1965).

34. V. E. Shashoua, J. Am. Chem. Soc. _82_, 5506 (1960).

35. D. S. McLure, J. Chem. Phys. 17, 905 (1949).

36. G. Porter and E. Strachan, Spectrochim. Acta 12, 299 (1958).

37. P. M. Johnson and A. C. Albrecht, J. Chem. Phys. 48, 851, (1968).

38. V. I. Stenberg in "Organic Photochemistry," Vol. I, Marcel Dekker, New York, N. Y., 1967, p. 127.

39. L. Schutte and E. Havinga, Tetrahedron 23, 2281 (1967).

40. J. C. Andersen and C. B. Reese, J. Chem. Soc. (1963), 1781.

41. H. Kobsa, J. Org. Chem. 27, 2293 (1962).

42. P. Adams and J. L. Anderson, J. Am. Chem. Soc. 72, 5154 (1950).

43. P. E. Stevenson, J. Mol. Spectry. 15, 220 (1965).

44. G. Porter and E. Strachan, Trans. Faraday Soc. 54, 1595 (1958).

45. S. P. McGlynn, T. Azumi and M. Kinoshita, "Molecular Spectroscopy of the Triplet State," Prentice-Hall, Englewood Cliffs, N. J., 1969, p. 159.

46. J. I. Dunlop, Photochem. and Photobiol. 5, 227 (1966).

47. D. B. Wetlaufer in "Advances in Protein Chemistry," C. B. Anfinsen, et al., Vol. 17, Academic Press, New York, N. Y., 1962, p. 303.

48. W. J. McCarthy and J. D. Winefordner in "Fluorescence, Theory, Instrumentation and Practice," G. G. Guilbault, ed., Marcel Dekker, Inc., New York, N. Y., 1967, p. 412.

49. R. A. Keller, in "Molecular Luminescence," E. C. Lim, ed., W. A. Benjamin, N. Y., 1969, p. 453.

50. G. Dobson and L. I. Grossweiner, Rad. Res. 23, 290 (1964).

51. L. Pauling, "Nature of the Chemical Bond," Cornell University Press, Ithaca, N. Y., 1948, p. 204.

52. W. A. Gibbons, G. Porter and M. I. Savadatti, Nature 206, 1355 (1965).

53. Yu. A. Vladimirov and E. E. Fesenko, Photochem. Photobiol. 8, 209 (1968).

54. A. Maybeck and J. H. Windle, Photochem. Photobiol. 10, 1 (1969).

55. K. Watanabe, J. Chem. Phys. 26, 542 (1957).

56. W. Gordy, W. B. Ard and H. Shields, Proc. Nat. Acad. Sci. U.S. 41, 983 (1955).

57. C. J. Seliskar, D. C. Turner et al., in "Molecular Luminescence," E. C. Lim, ed., W. A. Benjamin, N. Y. 1969, p. 677.

58. L. I. Grossweiner and W. A. Mulac, Rad. Res., 10, 515 (1959).

59. I. Fischer-Hjalmars and M. Sundbon, Acta Chem. Scand. 22, 607 (1968).

60. H. B. Steen, Photochem. Photobiol. 9, 479 (1969).

61. E. Yeargers and L. Augenstein, Biophys. J. 5, 690 (1965).

62. Yu. A. Vladimirov and E. A. Burstein, Biofizika 5, 385 (1960).

63. W. C. Price and A. D. Walsh, Proc. Royal Soc. A191, 22 (1947).